Research on E-Learning and ICT in Education

Ioannis Kazanidis · Avgoustos Tsinakos
Editors

Research on E-Learning and ICT in Education

Technological, Pedagogical and Instructional Perspectives

Springer

Editors
Ioannis Kazanidis
Department of Informatics
Democritus University of Thrace
Kavala, Greece

Avgoustos Tsinakos
Department of Informatics
Democritus University of Thrace
Kavala, Greece

ISBN 978-3-031-99960-4 ISBN 978-3-031-99961-1 (eBook)
https://doi.org/10.1007/978-3-031-99961-1

This Springer imprint is published by the registered company Springer Nature Switzerland AG
The registered company address is: Gewerbestrasse 11, 6330 Cham, Switzerland

About This Book

This edited volume presents current, peer-reviewed research on e-learning and ICT in education. It provides expanded peer-reviewed content from research presented at the 13th Pan-Hellenic and International Conference "ICT in Education", held in Greece in 2023. While many chapters are based on work presented at the conference, the volume also includes submissions from an open international call, attracting contributions from researchers beyond the conference community.

It offers a broad and practice-oriented perspective on how digital technologies are reshaping teaching, learning, and teacher development across diverse educational settings. It brings together contributions from researchers and academics responding to an international call for chapters, with selected works reflecting both conceptual developments and applied innovations grounded in real-world educational contexts.

The volume includes studies spanning early childhood, primary, secondary, and higher education, as well as teacher training and instructional design. Emphasis is placed on methodological diversity and relevance to educational practice, with several chapters addressing design, implementation, and evaluation of ICT-supported interventions. The research presented responds to current challenges in digital competence development, curriculum innovation, inclusive education, and the integration of advanced technologies such as artificial intelligence, serious games, and augmented reality.

The book is structured into three thematic parts to provide an integrated and coherent examination of the field.

Part, "*Teacher Education, Digital Competence, and Pedagogical Practices,*" explores the evolving role of teachers in technology-enhanced learning environments. It includes studies on computational thinking in early childhood, digital competence frameworks, innovative instructional approaches such as flipped learning, and factors influencing teachers' adoption of emerging technologies like big data.

Part, "*Instructional Design, Artificial Intelligence, and Serious Games,*" addresses the intersection of pedagogical design and technological advancement, with chapters focusing on AI applications in instructional design, taxonomies for support in serious games, and the role of game-based learning in higher education and teacher preparation.

Part, “*Inclusive and Immersive Technologies in Education,*” investigates the application of AR and robotics in inclusive contexts, highlighting the potential of immersive technologies to support learners with disabilities and promote engagement through innovative educational scenarios such as disaster management simulations.

This volume serves as a comprehensive reference for researchers, educators, instructional designers, and policymakers who are engaged in the design, study, or implementation of technology-enhanced learning strategies and digital transformation in education.

Contents

About the Editors

Ioannis Kazanidis is an Assistant Professor at the Department of Informatics of the Democritus University of Thrace, Greece. He is the Director of the international M.Sc. in Immersive Technologies—Innovation in Education, Training, and Game Design and serves as the Vice Director of the Advanced Educational Technologies and Mobile Applications Lab at the same university. He has a BSc from the Department of Mechanical Engineering Educators from the School of Pedagogical and Technological Education (ASPAITE), B.Sc. from the Department of Informatics from the Hellenic Open University, M.Sc. in Computing from Coventry University and Ph.D. in Educational Technology from the Department of Informatics of University of Macedonia. He has been actively involved in several internationally funded research projects focusing on immersive technologies in education and educational technology. His research interests include Technology-Enhanced Learning, Game-Based Learning, AR/VR in education, mixed reality environments, AI in education, learning analytics, and educational tangible user interfaces. He has authored more than 150 publications in international journals and conferences, with his work receiving over 3,500 citations.

Dr. Avgoustos Tsinakos is a Professor and Director of Advanced Educational Technologies and Mobile Applications (AETMA) Lab in the Department of Computing Science at Democritus University of Thrace and former Director of the International M.Sc. in Immersive Technologies in Education Training and Game Design.

He was/is Coordinator (or member of the Coordination Team) in more than 32 national/international programs related to Immersive Technologies, Distance Education and Technology Enhanced Learning/Moreover he collaborates scientifically with several Universities in USA, Spain, Canada, the Netherlands and China but also with government bodies such as the Prosecution of Electronic Crime in Greece. His publication list includes four books in: English and one in Chinese, four book chapters, more than 100 papers on the topic of e-learning/mobile learning and of Augmented Reality. In addition, Prof. Tsinakos has participated in more than 80 world/international/national conferences as invited speaker/presenter.

Teacher Education, Digital Competence, and Pedagogical Practices

Computational Thinking in Early Childhood Education. Significance, Approaches, and Challenges Ahead

Maria Tsapara and Tharrenos Bratitsis

Abstract Computational Thinking (CT) is an essential twenty-first-century competence focusing on systematic problem-solving and supporting interdisciplinary learning. This study presents a systematic literature review examining the integration of CT in Early Childhood Education (ECE), underlining its core components—decomposition, pattern recognition, abstraction, and algorithmic thinking and emphasizing on innovative pedagogical strategies, including the use of educational robotics and unplugged activities. Tools like Bee-Bot, Blue-Bot, mTiny, and KIBO cultivate children's problem-solving skills, critical thinking, and creativity, while unplugged activities, such as board games, storytelling, and other playful activities, support cognitive and motor skills development. The connection between Board games and CT fosters interactive and hands-on learning while enhancing collaboration. Challenges include curricula integration and professional development of educators, supporting them to understand how to use technology in an effective way. This systematic review reinforces the vital role of CT to nurture creativity, digital literacy, and interdisciplinary learning in early years, aligning with global educational standards and preparing children for a tech-driven future.

Keywords Computational thinking · Early childhood education · pedagogical strategies

1 Introduction

Computational Thinking (CT) is recognized as a core competence required in the twenty-first century [1], while it has been promoted as a competence that is as fundamental as numeracy and literacy [2]. This systematic review provides an overview of

M. Tsapara (✉) · T. Bratitsis
Early Childhood Education Department, University of Western Macedonia, Florina, Greece
e-mail: m.tsapara@hotmail.com

T. Bratitsis
e-mail: bratitsis@uowm.gr

I. Kazanidis and A. Tsinakos (eds.), *Research on E-Learning and ICT in Education*,
https://doi.org/10.1007/978-3-031-99961-1_1

the significance of CT in Early Childhood Education (ECE), as it plays a crucial role in fostering critical thinking and problem-solving skills for young learners. Introducing CT concepts in early years cultivates curiosity and prepares children for the complexities of the digital future. The review further explores the potential impact of strategies and tools and their practical applications in the classroom. Innovative pedagogical approaches, such as robotics, unplugged activities, and gamified learning, are the key to enhancing young children's engagement with CT. Additionally, the study underlines challenges including the necessity for teacher continual professional development and curriculum refinement while aiming to provide a thorough understanding of how CT can be effectively integrated into ECE. This approach ensures that all children—regardless of background—are equipped with the vital skills needed to flourish in a technological era.

CT is a way of thinking used to solve problems that can be comprehended and performed by a computer. Many researchers highlighted the significance of CT as an essential skill for everyone and remarked that CT transversally applies creativity in and beyond computing disciplines as it offers a different way of looking at and resolving problems that can be computed [2, 3]. Papert [4] set the foundation for introducing CT into the educational process and Wing [3] advocated for this incorporation, while both highlighted how computers may be used as tools to improve creativity, critical thinking, problem-solving and help people in solving problems more efficiently, regardless of their field of study.

The International Society for Technology in Education (ISTE) and the Computer Science Teachers Association (CSTA) pointed out that "*CT is a problem-solving process that includes (but is not limited to) the following characteristics: (1) Formulating problems in a way that enables us to use a computer and other tools to help solve them, (2) Logically organizing and analyzing data, (3) Representing data through abstractions such as models and simulations, (4) Automating solutions through algorithmic thinking (a series of ordered steps), (5) Identifying, analyzing, and implementing possible solutions with the goal of achieving the most efficient and effective combination of steps and resources and (6) Generalizing and transferring this problem-solving process to a wide variety of problems*" [5].

Brennan and Resnick [6] examined CT as more than just a conceptual approach to problem-solving. They provide an extensive overview of CT by examining concepts, practices, and perspectives, which may be used as a framework for thinking more deeply and broadly about the world in a computational way as well as a tool for solving specific problems.

According to Wing [3] there are four key components of CT: Decomposition refers to breaking down complex problems into smaller components, focusing on the parts and solving them step by step. This makes it easier to solve each part of the problem individually [7, 8]. Pattern recognition applies to identifying patterns and similarities in problems and optimizing the problem-solving process to become more effective [9, 10]. Abstraction relates to the process of selecting, filtering, and hiding unnecessary and irrelevant information/details of a problem and focusing only on its essential aspects to develop a generalized solution [11]. Algorithm design involves determining a sequence of steps to solve a problem, applying instructions in the correct order, and repeating this sequence multiple times. This process ensures accuracy as algorithms are used to solve problems or carry out tasks in a repeatable and efficient way [12].

Kalelioğlu et al. [13] stated that CT used to be exclusive to particular age groups and disciplines; however, it is now widely acknowledged as a vital skill for people of all ages and in every field. The ubiquitous integration of technology, rapid technological advances in computer efficiency, and the numerous advantages of living in the digital era are all contributing significantly to this change. Meanwhile, developing CT skills affords people an essential advantage in problem-solving by enabling them to handle real-life situations with greater effectiveness.

Debugging is another essential concept in programming and CT. An important aspect of the CT process is making mistakes, identifying them, and debugging [8]. After incorporating several approaches to formulate the construct in K–16 settings, Shute et al. [14] defined CT as "The conceptual foundation required solving problems effectively and efficiently with solutions that are reusable in different contexts". They also demonstrate iteration ("Repeat design processes to refine solutions, until the ideal result is achieved") and generalization ("transfer CT skills to a wide range of situations/domains to solve problems effectively and efficiently").

According to Grover and Pea [8], CT includes core concepts that are beneficial in problem-solving and contributes to the development of digital literacy, which has emerged as an essential skill in many fields. Researchers also pointed out that abstraction, pattern recognition, representations, algorithmic design, decomposition, and iteration are among the fundamental components of CT. Conditional logic, parallel thinking, debugging, and efficiency considerations are additional components of CT. These constitute the basis of curricula aimed to foster and evaluate CT skills, providing students the skills they require for addressing challenging, real-world problems.

Wing [15] noted that "Our immediate task ahead is to better explain to non-computer scientists what we mean by CT and the benefits of being able to think computationally", these principles are not just for computer scientists; they are beneficial for anyone solving complex problems, regardless of their expertise. The integration of these components promotes clearer thinking and more structured problem-solving strategies. The National Research Council [16] emphasized that "(1) students can learn thinking strategies such as CT as they study a discipline, (2) teachers and curricula can model these strategies for students, and (3) appropriate guidance can enable students to learn to use these strategies independently".

Wing [17] also stressed that implementing CT in educational systems would help students and teachers understand how to approach problems methodically, breaking them down into smaller and solvable parts while using algorithms to find solutions. Rich and Langton [18] revealed a commonly agreed-upon set of foundational concepts and processes for CT. These could be used as a guide for what students need to know when they are first introduced to computing in early learning initiatives.

Barr and Stephenson [12] noted that it is crucial to foster CT in K–12 education [19]. According to their research, CT promotes confidence in dealing complexity, perseverance in resolving challenging problems, ability to lead, and the capacity to deal with open-ended challenges [20]. If our goal is to nurture a generation with CT skills, we need to familiarize students with CT in K-12 education at early ages.

Despite the growing recognition of CT as a fundamental skill, its integration into ECE remains underexplored. While extensive research has been conducted on CT in K-12 and higher education [12, 19], limited studies focus specifically on how CT concepts can be effectively introduced in early learning environments. Many existing studies primarily address CT within the context of computer science education, leaving a gap in understanding its interdisciplinary applications in early childhood settings [13]. This systematic literature review aims to bridge this gap by exploring innovative pedagogical strategies—including robotics, unplugged activities, and gamified learning—that can enhance young children's engagement with CT. Additionally, it examines the challenges of integrating CT into ECE, such as the need for continuous teacher professional development, curriculum refinement, and the accessibility of age-appropriate CT resources. This review aims to address the following research questions:

1. How is Computational Thinking (CT) integrated into ECE?
2. How do young children engage with CT concepts such as algorithmic thinking, pattern recognition, abstraction, decomposition, and debugging?
3. What are the most effective strategies and tools for integrating CT in early childhood education?
4. What challenges exist in implementing CT activities in early education, and how can they be addressed?

To explore these questions, a systematic review will be conducted, analyzing evidence-based research and initiatives from the past decade that focus on CT in ECE. By synthesizing existing studies, this review provides a comprehensive overview of the prevailing practices and experiences that define the integration of CT in ECE.

2 Methodology

A systematic literature review (SLR) was conducted to synthesize research findings and identify gaps in the integration of CT in ECE [21]. This review systematically collects, evaluates, and synthesizes relevant studies from online and printed sources, including books and peer-reviewed journal articles. The primary objectives are to establish the current state of knowledge on CT in ECE, identify the most effective pedagogical strategies (e.g., robotics, unplugged activities, gamified learning), and highlight areas requiring further investigation. By following a structured methodology, this review ensures a comprehensive and unbiased analysis of existing research in the field.

2.1 Literature Search Process

The study employed a systematic literature review (SLR) design, which was carried out in two main stages: the identification and selection of relevant literature, followed by data extraction and synthesis. This methodology aimed to systematically address the research questions and analyze existing studies to draw conclusions about the integration of CT in ECE.

Stage 1: Literature Identification and Selection The first phase of the research focused on identifying relevant literature through a systematic search process across multiple academic databases. The aim was to locate studies that address the integration of CT into ECE and provide insights into how children engage with CT concepts such as algorithmic thinking, pattern recognition, abstraction, decomposition, and debugging.

Database Search The following academic databases were systematically searched for relevant literature: Scopus, ERIC, and Google Scholar. These databases were selected due to their comprehensive coverage of education-related research, including peer-reviewed journals, conference proceedings, and dissertations.

Search Terms and Keywords A combination of the following keywords was used in various combinations to ensure the breadth of the search. This search string ensures that all key concepts are addressed, including the specific CT skills (e.g., algorithmic thinking, pattern recognition) and methods/tools (e.g., unplugged activities, robotics), while also accounting for integration strategies, challenges, and empirical studies.

("computational thinking" OR "CT") AND ("early childhood education" OR "ECE") AND ("unplugged activities" OR "robotics in education" OR "gamified learning") AND ("algorithmic thinking" OR "pattern recognition" OR "abstraction in ECE" OR "decomposition in ECE" OR "debugging in early childhood") AND ("integration" OR "implementation" OR "strategies" OR "tools" OR "activities") AND ("challenges" OR "barriers" OR "effectiveness") AND ("empirical study" OR "intervention" OR "assessment")

Data Screening Process The first screening involved reviewing the titles and abstracts of the articles retrieved from the databases. Articles that did not meet the criteria were excluded at this stage (Table 1).

The second round of screening involved a full-text review of the remaining articles to ensure they met all inclusion criteria. Only those that directly addressed the integration of CT in ECE were selected for the review.

Stage 2: Data Extraction and Synthesis Once the relevant articles were identified and selected, the second phase of the research focused on data extraction and synthesis of the findings. During the data extraction process, key information was gathered from each selected study to ensure a comprehensive understanding of the

Table 1 Inclusion and exclusion criteria table

Criteria	Inclusion criteria	Exclusion criteria
Article title and content	Article title is relevant and aligns with the study's focus on CT in ECE	Article title does not match the study's requirements and has irrelevant or unrelated content
Focus on age group	Studies involving children aged 4–8 years, focusing on early childhood education (ECE)	Studies focused on higher education or elementary education (ages 8 and above)
Year of publication	Publications from 2012 to 2024	Publications outside the range of 2012–2024
Type of publication	Solely journal articles that provide empirical data and align with the study's criteria	Reviews, editorials, and non-empirical studies (e.g., opinion papers)
Language	Studies published in English	Studies published in languages other than English
Accessibility	Full-text articles or open-access articles	Articles that require payment or cannot be accessed in full

integration of CT in ECE. The extracted data included the study design, identifying whether the research was experimental, quasi-experimental, a case study, or a literature review. Additionally, the CT concepts covered in each study were noted, with particular attention to how algorithmic thinking, pattern recognition, abstraction, decomposition, and debugging were addressed. Information was also collected on the strategies and tools used for integrating CT in ECE, such as unplugged activities, robotics, and gamified learning. The educational context, including the specific settings and age groups involved (preschools, kindergartens, and early education centers), was another important data point. Finally, the outcomes and implications were extracted, focusing on the effectiveness of CT integration and any challenges identified.

The synthesized data was categorized into thematic areas based on the research questions. The first thematic area, *Integration of CT in ECE*, explored how CT was being incorporated into early childhood classrooms, both through structured curricula and informal activities. The second area, *The Importance of Early Education in Developing Computational Skills*, provided insights into how young children engage with and develop CT skills, emphasizing the specific CT concepts such as algorithmic thinking, pattern recognition, abstraction, decomposition, and debugging. The third thematic area, *Tools for Introducing Computational Thinking in Early Child-hood Education*, analyzed the most successful pedagogical strategies, tools, and interventions for teaching CT in early childhood settings. The final theme, *The Future of Computational Thinking in Early Education: Conclusion and Critical Reflections*, focused on understanding the difficulties educators face when implementing CT and how these challenges can be addressed, especially in diverse ECE environments. This data synthesis helped to build a holistic view of the current landscape of CT integration in ECE.

3 The Importance of Early Education in Developing Computational Skills

Children in early education need to be computationally literate [22], while at the same time they can learn to address real-world problems by using systematic approaches [23]. They are also able to understand the concept of algorithms by using examples from the real world. Furthermore, by cultivating their algorithmic skills, they develop thinking strategies to understand and solve problems at a very early age [24].

Algorithmic thinking, a key CT component, aims to analyze computational problems while defining algorithms as the framework for solutions before they are implemented [3]. According to Mezak et al. [24], it "*refers to the ability to break down problems into smaller, manageable tasks, which can be systematically solved*". Cultivating algorithmic thinking in (ECE) needs developmentally appropriate tools while the introduction of these concepts must align with children's natural learning styles, which often involve play, inquiry, and exploration.

Wong et al. [25] demonstrated that children who used block-based programming tools improved their CT skills and specifically algorithmic thinking, their ability to recognize problems and come up with structured solutions. Masarwa et al. [26] incorporated in their study a pretest-intervention-posttest design to assess changes in the CT skills of 16 kindergarteners. A learning unit with card games was developed, focusing on two fundamental sorting algorithms: Linear and category sort. Their study shares critical knowledge into how specific algorithmic concepts can be taught effectively in early education. Vujičić et al. [27] examined ways of incorporating algorithmic thinking skills into everyday activities in kindergarten, underscoring the role of educators in cultivating these skills.

According to Acosta et al. [28], designing algorithms and recognizing patterns are two CT key skills that can potentially be introduced early in childhood education. Piaget's [29] theory of cognitive development forms the basis for understanding how children learn and recognize patterns in their environment [30]. Pattern recognition involves recognizing patterns, similarities, or the unique characteristics of a problem [31, 32] to anticipate results or discover resolutions; this aspect of CT can enhance the problem-solving process and manage complex difficulties efficiently [10]. Pattern identification involves the observation and recognition of regularities or iterative sequences in objects or data [33], while the knowledge of repeating patterns is a predictor of broader mathematical abilities, such as counting and understanding numerical principles [34]. These principles are also identified in CT with similar concepts such as iterative loops and sequences, which are part of programming and algorithm design.

Acosta et al. [28] carried out an extended analysis of repetition pattern understanding. They investigated how children aged 3–5 understand and justify repetition patterns within technological contexts, bridging algebra and CT. A mixed study revealed developmental progress in children's pattern representation and reasoning processes. Calderon et al. [35] studied the development of CT through pattern recognition in early years. They focused on UK's educational reforms, underscoring digital skills and CT in school curricula, and explored their implementation in ECE. Saxena et al. [36] stated that young children can grasp structured problem-solving when instruction is age-appropriate, while sequencing and pattern recognition can successfully be taught.

Promoting the teaching of pattern recognition in ECE contributes to the development of mathematical thinking, particularly to the process of representation and abstraction in mathematics [28, 37, 38]. According to Wing [39] abstraction is one of the 11 thinking processes that define CT, where "*this process—deciding what details we need to highlight and what details we can ignore—underlies CT*". She also stressed that CT is more than just using computers; it's a way of thinking and motivates students to think in a logical way and break down complex problems when they are asked to apply multiple layers of abstract thinking.

By focusing on abstraction, structured activities can not only support children in the problem-solving process but also improve their broader cognitive abilities and encourage them to engage in abstract thinking through exploration and innovation [40].

Qian and Choi [41] focused on the significance of abstraction in CT, especially within the context of K-12 education and STEM literacy. They also identify three cognitive processes after reviewing cognitive dimensions of abstraction and the role of abstraction in computing education: (1) Filtering information, (2) Locating similarities, and (3) Mapping problem structures. Their study indicated the need to address existing educational gaps by supplying a defined framework and powerful learning methods. Acosta et al. [28] argue that students can understand abstract concepts through technological environments in a more playful way, such as online games, animation programming for children, and educational robots [42]. By using robots, students can be taught patterns, helping them understand abstract concepts in a more efficient way [43].

Kramer's findings [44] indicate that about 50% of the conceptual definitions considered abstraction solely as a form of decomposition activity. Ezeamuzie et al. [44] identified the main activities of decomposition as: (a) identifying relevant and irrelevant details of an entity, (b) obscuring or removing irrelevant details, and (c) focusing on only the important details. According to Shute et al. [45], decomposition involves breaking down a complex problem into smaller, manageable parts to allow for a more convenient solution [15, 46]. Moschella [47] states that studies focusing exclusively on decomposition in early childhood are relatively scarce; as it is regarded as both a CT skill as well as a problem-solving skill [3].

Selby [48] argues that decomposition is recognized as one of the most difficult CT skills to acquire. One contributing factor to this challenge is that sometimes the problem to solve isn't completely understood. Furthermore, he states that students often encounter difficulties when attempting to apply the concept of decomposition in unfamiliar contexts, despite having an understanding of it. Dietz et al. [49] investigated the early development of problem decomposition abilities in children to determine their value in cognitive development and implications for educational strategies. Cheng [48] states that decomposition in relation to arithmetic problems is already taught to students as young as 5 years old, who have to invest in more lessons on decomposition in relation to CT in early primary school. The findings by Dietz et al. [49] highlight the foundational role of decomposition in cognitive development and its implications for educational strategies. According to Zeng et al. [50], CT practice of decomposition can be learned or applied in children's daily routines, such as getting ready for school, washing hands, toileting, and making bread. Ehsan et al. [51] indicated that children aged 7–8 years decomposed the problem of building a space for a puppy into smaller parts. This process enabled them to identify the main parts of the task along with the essential criteria necessary for successful completion.

Angeli & Valanides [52] showed that the majority of the children's problem-solving strategies during debugging used decomposition as a strategy to deal with the complexity of the task. They highlight that "*these results are important, because they show that children at this very young age are able to cope with the complexity of a learning task by decomposing it into a number of subtasks that are easier for them to tackle*". Researchers stress the need to nurture these skills in early years, as they significantly contribute to cultivating critical thinking and problem-solving skills necessary for addressing complex problems in various contexts [51, 53].

As children gain the ability to break down complex spatial problems into manageable parts, they are involved in a critical cognitive process that parallels the debugging skills discussed by Deliema et al. [54]. They define debugging as "*The process of handling breakdowns in computer programming*" or "*fixing syntactical errors in a program*" [55]. Sun et al. [56] highlighted that it is "*an integral part of CT skills when a solution plan does not work as expected*". Debugging has been also defined as the process of identifying and correcting errors in early childhood, while researchers state that these skills can be enhanced through purposeful instruction [57] and a high-level thinking skill that is a core component and a critical aspect related to CT development in early years [58]. Sun et al. [56] argue that debugging is particularly underrepresented in the CT training literature and especially in the context of ECE. Usually, CT concepts such as pattern recognition and sequencing are often emphasized, debugging receives less attention [59].

Bers et al. [60] underline the effectiveness of the TangibleK curriculum in teaching CT, programming concepts and robotics to 53 children of 3 kindergartens. Debugging skills involved recognizing problems, setting goals, identifying potential reasons or explanations for a problem, and testing solutions. Cultivating debugging skills reflects the Engineering Design Process, highlighting the value of problem-solving, and acknowledging that failure is part of the learning process while the results show that children can effectively engage with and apply these concepts [60].

Misirli and Komis [58] investigated the debugging process of 526 preschool children (4–6 y.o.) while programming a tangible robot. Furthermore, they examined the strategies that children use to debug effectively, the kinds of mistakes that occur when children program the tangible robot, and how they can recognize errors. The main finding was that, through debugging, children developed an understanding of coding rules and structure (syntactic knowledge) and gained insight into the purpose of coding (semantic knowledge). Their research provides valuable insights into how children build foundational programming knowledge and engage in debugging, contributing to the broader understanding of nurturing CT skills from an early age.

Pugnali et al. [55] underscored that debugging is recognized as the most complex concept within CT skills. Using tangible robotic kits like KIBO enhances debugging skills and hands-on learning, rather than through visual-block programming platforms such as ScratchJr. They also pointed out that it promotes positive socio-emotional behaviors, enhancing overall learning experiences.

Silvis et al. [61] conducted a design-based study that focused on debugging as a learning process. They also aimed to elaborate how children encounter and resolve errors while operating coding robots. The study focused on collaborative debugging activities where children worked in groups to repair bugs in various coding programs. According to Silvis et al. [61], integrating debugging as a critical educational tool in ECE is essential for teaching coding and robotics effectively.

Bers [62] states that there are seven "powerful ideas", core concepts of the CT framework that are ideal for early childhood development. These are: algorithms, modularity, control structures, representation, hardware/software, design process, and debugging.

According to Bers et al. [63], these "powerful ideas" can be related to early childhood foundational skills (Table 2).

Cultivating CT in ECE is critical, as it enables children to transition from passive users to active creators of technology [22]. CT equips children with the tools to address real-world problems by enhancing skills such as abstraction, algorithmic thinking, debugging, decomposition, and pattern recognition and through systematic

Table 2 Connecting CT concepts with foundational skills in early childhood [63]

Computational thinking concepts	Early childhood foundational skills
Algorithms—a series of ordered steps taken in sequence	Order matters Logical organization
Modularity—breaking down complex tasks and procedures into simpler, manageable units	Breaking down a large task
Control structures—controlling the sequence in which a program is executed. Making decisions based on conditions	Pattern recognition
Representation—concepts can be represented by symbols	Symbolic representation of letters and numbers
Hardware/software systems—computing systems need both hardware and software to operate	Recognizing objects and processes that are human engineered
Design process—an iterative process used to develop programs and artifacts with multiple steps	Writing process Scientific method Creative process
Debugging—fixing problems in our programs in a systematic way	Perseverance Problem-solving

pedagogical approaches and strategies [23]. Researchers indicate that integrating developmentally appropriate tools, such as unplugged activities, tangible robotics, and playful educational strategies, enhances CT skills in early years [25, 60]. By introducing these core components of CT in ECE, researchers advocate for a holistic approach that aligns with innovation, inquiry-based learning and emphasizes play, by matching children's natural learning styles [64].

4 Tools for Introducing Computational Thinking in Early Childhood Education

4.1 Introducing Robotics: A Hands-on Approach for Kindergarten

Robotics is a tool that early childhood educators use to introduce CT in their classrooms. It helps children to reflect on their ideas [65] and cultivate twenty-first century skills. It also supports the development of CT by focusing on parameters such as algorithms, debugging, pattern recognition, and deconstruction [66]. It engages students in hands-on activities that make learning about technology, science, and programming a fun and interactive experience [67, 68].

In recent years, a number of studies have assessed different aspects of educational robotics and programming in ECE, emphasizing their effects on skill growth and learning outcomes. Research has demonstrated that there are many benefits of integrating robotic technologies in early years [69]. Zviel-Girshin et al. [67] state that young learners feel they can write a new program or code a new behavior of a robot, while they can read and explain the code of their final project. According to Sullivan et al. [69], manipulating robotic objects supports children in developing fine motor skills [70]. Researchers state that utilizing educational robots can foster CT [71] and critical thinking [72], enhance motivation for learning, improve social skills, collaboration [73] and communication [74], empower creative expression [69], introduce technology, and support differentiated learning for diverse learners.

Over the past few decades, many educational robots have been released that are age-appropriate for ECE. In the 1980s, the Logo Turtle was introduced as one of the first educational robots, used in ECE as part of the Logo programming language initiative [4]. The "Roamer robot" was created in the 1990s by Professor Tom Stonier in the UK and used in ECE, introducing young children to concepts of movement and control [75].

Studies highlight that the programmable floor robots Bee-Bot and Blue-Bot are developmentally appropriate and easy-to-use tools for preschoolers. These robots can be integrated into the educational process through playful activities [76, 77] and enhance several skills such as problem-solving, algorithmic thinking, critical thinking, spatial orientation, debugging, programming, logical thinking [78, 79], and spatial thinking [80]. At the same time, it can support interdisciplinary learning [81].

Numerous articles and studies have been published regarding KIBO's impact on ECE, including works by Pinto et al. [82], Elkin et al. [83], and Sullivan and Bers [42], which explore the integration of robotics into preschool curricula and its effects on CT skills. Mikova and Fejková [84] integrated two programmable robots, mTiny Coding Kit and Tale Robot—Matatalab. The authors analyze these kits with regard to programming concepts, interactivity, technical security, their applicability in educational contexts, and their effectiveness in introducing basic algorithmic thinking. Değirmenci [85] focused in coding through two approaches: one using the mTiny robot and another using activity-based coding with non-technological materials. Both methods significantly improved cognitive flexibility, while activity-based coding showed greater gains in CT, highlighting the value of diverse strategies in early childhood coding education. Focusing on utilizing mTiny Coding Kit in kindergarten, Di Donato and Mattioli [86] stress that "this robot is really engaging for students because the robot has a screen where it shows the eyes of the Panda making visible the emotions it feels. This robot is a first step to talk about emotion and robotics".

In the framework of the project CThink.IT: Think, Learn, and Play in a CT Way, co-funded by the European Union, Erasmus+ Program, Key Action KA2, School Education, Busuttil et al. [66] reviewed 51 papers on robotics approaches to CT from 2012 to 2023. This review underlines the growing popularity of using robotics to teach CT in ECE. Moreover, of the total 51 papers that discuss the use of robotics in the context of CT, 35 papers focus on robotics implementations of CT, 11 papers combine robotics with unplugged activities, 3 papers combine robotics with programming, and only 2 papers explore the combination of all three methods. Through their research various robots were found to be implemented to teach CT such as Bee-Bot, KIBO, Cubetto, Blue-Bot, Matatalab/tale robot, RoboTito, Social Robot (SoRo) Toolkit, ALERT & Robopad, Botley, Code-a-pillar, CodeAttach, Codey Rocky, Cubelets, iRobiQ, Lego Mindstorms NXT, Lego WeDo, LinkBricks, Mouse Activity, Pioneer 3-DX, Sapling, Sphero, TACTOPI, Tangibot, Thymio, TurtleBot, U-Bot.

Integrating robotics in ECE, it is equally important to implement unplugged activities, which provide opportunities for children to develop CT, enhance creativity and cognitive skills without relying on technology [87].

4.2 Fostering Computational Thinking through Unplugged Activities

CT is recognized as an essential skill for ECE because it supports children's cognitive development [22, 41, 54]. It is also important that children go beyond merely using technology; they should try to understand how it works and how to apply it to solve real-world problems. This approach fosters creativity [88, 89] develops algorithmic thinking [12], and enhances problem-solving skills [9]. Romero et al.

[90] in their study analyzed unplugged activities from the point of view of embodied cognition highlighting the advantages and challenges of using robots in CT activities, comparing and contrasting them with unplugged approaches.

Recent findings stress the value of unplugged activities in fostering CT skills among early learners [87, 91, 92]. Unplugged activities are educational methods that teach CT through hands-on, physical activities teaching computer science (CS) and programming concepts without using electronic devices such as computers or tablets [19, 93]. These activities introduce children to essential ideas such as debugging, algorithm development, problem-solving [73, 94], coding, command sequences, and repetition through hands-on, playful experiences. Chen et al. [95] conducted a systematic review and meta-analysis to understand the potential of unplugged activities on CT skills. This review examined 49 studies conducted between 2006 and 2022, highlighting the positive impact of unplugged activities on students' CT skills across K-12 education. Küçükkara and Aksüt [96] emphasize the value of unplugged coding activities for kindergarten students, allowing them to engage with core components of CT without relying on computers or technology.

Mladenovic et al. [97] examined the utility of CS-unplugged activities in teaching fundamental concepts to children aged 3–6 years. Their study emphasized multiple advantages of introducing CT through unplugged activities, including measurable improvements in problem-solving skills, enhanced algorithmic thinking skills and early exposure to core computational principles. These benefits were assessed through observational studies and pre- and post-activity evaluations. Çakıroğlu et al. [98] focused on the cultivation of problem-solving skills through unplugged activities, offering meaningful perspectives of improvement in ECE. The analysis focused on 11 children aged 4–5 in a public preschool, using Code.org activities as an instructional package. Results showed that CS-unplugged activities positively enhanced children's understanding and planning skills.

Unplugged activities could include directional coding with grids [99, 100], constructing a board game [101, 102], designing a virtual-physical dance performance [103], origami coding [104], direction game with cards [105], storytelling [106, 107], Tic-Tac-Toe and Sequencing stories [36], providing children with concrete experiences before transitioning to plugged-in virtual experiences with coding [99]. Robinson [94] investigates the impact of unplugged learning in teaching CS to kindergarteners. In her study children were taught using either strictly unplugged methods using games, books, manipulatives, and movement or a plugged/ online curriculum using online puzzles and videos, including some unplugged lessons.

Özdinç and Kandemir-Özdinç [108] highlight the contribution of unplugged CS activities in the development of motor and cognitive skills in early years. They used the BEST engineering design process by NASA. The activity, which focused on environmental cleanliness, was conducted on a programming carpet divided into squares. It involved 5-year-old children working collaboratively in groups. They found that the activity helped children improve their learning skills in a playful and engaged way while the CS activities supported CT and focused on all aspects of a child's growth. Serving as a way to introduce CS concepts in ECE, Wang et al. [109] underlined the significance of gamified approaches in children's coding education which proved to gain support from kindergarten and elementary school teachers. Furthermore, they demonstrate the effectiveness of integrating gamification, unplugging, and cost-efficiency as design principles.

Moreover Masarwa et al. [26] designed a learning unit based on card games that teach two sorting algorithms: Linear Sort and Category Sort. Sixteen kindergarten children participated. A pretest-intervention-posttest single group design was employed to measure CT skill development. Children improved their CT skills, progressing from an intermediate to a sophisticated level. In addition, Saxena et al. [36] state that through unplugged activities and by using tangible materials children have concrete experiences with CT concepts such as pattern recognition, sequencing, and algorithm design while preparing them with the necessary vocabulary and instructions for plugged activities. In their study, students were required to arrange several pictures of everyday events and daily routines. Through this unplugged activity students cultivate their pattern recognition and sequencing skills, preparing them to design a path with its corresponding sequence of commands in the Bee-Bot activity.

The integration of gamified approaches, including card games and unplugged activities, points out the increasing awareness of CT as a fundamental skill in ECE. This is in tandem with other trends such as the use of board games which are also playful and hands-on techniques, to clarify complicated concepts and enhance key skills from early childhood.

4.3 Board Games: Building Computational Thinking Skills through Play

The significance of CT in education is continuously growing, and many view it as "the new literacy of the twenty-first century" [3]. In this framework, along with the growing trend of gamification in education, board game-based learning is also associated with ECE [102]. Bayeck [110], states that "*board games simplify complex issues and systems, which make them appropriate to further explore learning and concepts such as motivation and CT in formal and informal settings*". According to Ching et al. [111], "*CT has been promoted as an important and fundamental skill for learners of all ages*". Researchers and educators tried to find out strategies to cultivate CT, in a more playful way such as board games [112, 113].

Board games encourage players to set and learn the rules, make decisions, be patient, predict outcomes through the selection of alternative moves, strengthen their relationships with others, and learn from experience [114, 115]. Board games aim to introduce students to concepts related to CT such as deconstruction, algorithms, patterns, problem-solving, debugging, iteration, and generalization [116].

Scirea and Valente [117] in their study explored board games and examined the connections between game mechanics, play styles, and other game elements with CT concepts and practices. Their effort led to a detailed analysis of selected board games and the development of a method to help educators and game designers identify the CT-supporting qualities of existing board games.

The philosophy of board games shares several similarities with competencies related to CT, as it promotes systematic problem-solving, setting goals and developing strategies. Similar to CT, where problems are examined and appropriate solutions are developed, players think creatively while using their imagination to consider different options and learn how to get beyond obstacles they encounter [88, 89]. Furthermore, in board games, just as players set their goals and analyze the strategies they will follow in CT, one needs to define the problem, develop, and choose the best solution [118, 119]. Assembling CT, whereby exchanging ideas and thinking together frequently results in more creative solutions [88, 120], in many board games fosters collaboration and teamwork among participants.

Apostolellis et al. [112] chose the operational definition of CT from the International Society for Technology in Education (ISTE), especially for K-12 education, due to its relevance and precision to their target audience. This defines CT as a way of solving problems which it includes automating solutions through algorithmic reasoning. The viewpoint of the algorithm defines the variables in the problem and the solution as a set of ordered action chains to be executed for solving the problem.

As the board game philosophy appears to have similarities with CT competence, exploring this correlation in early years is still not extended. Based on this research gap, the study by Tsapara et al. [102] focused on a literature review of board games that have been utilized in the classroom to cultivate CT skills. They examined eight board games, all of which aimed to solve a problem through a playful process while also fostering CT components such as decomposition, abstraction, algorithmic design, and pattern recognition. Additionally, key design elements of board games were documented, such as the game board, the use of cards and tokens, as well as significant mechanisms like game rules, card collection or discarding, action points, movement points, pattern movement and action.

Taking into account the wide range of benefits of integrating board games and CT activities into the educational process in ECE, it becomes evident that further research should focus on the design and utilization of board games that foster CT components in early childhood, Bratitsis et al. [101] created a board game called "Mystery at the Museum", where the players had to use basic competencies of CT. The initial results indicate that the board game shows potential as a useful tool for cultivating CT skills. Addressing the identified problems and implementing the suggested improvements could greatly increase the game's educational worth and practicality in classroom environments.

Uribe & Cobos [121] stressed that integrating hands-on activities with thorough knowledge, board games appear to be an ideal match for stimulating children's creativity, imagination, and curiosity. Tsai et al. [122] claim that board games allow students to learn from each other, cultivating interactivity and student-centered learning. Board games provide a variety of mechanics that provide an essential and versatile tool for developing CT skills in educational environments [66, 123]. Additionally, they include feature elements such as goals, rules, collaboration, hands-on experiences (e.g., through pawn movement or using cards), turn-taking, interaction, competition, and problem-solving. Consequently, implementing board games in an educational setting can enhance and cultivate the growth of a range of skills and competencies [124].

5 The Future of Computational Thinking in Early Education: Conclusion and Critical Reflections

In recent years, the cultivation of CT in ECE has been recognized on a global scale, with various countries developing frameworks that include CT as part of their educational standards (Table 3). According to Bers et al. [63] "*There is a growing demand for countries to incorporate some form of computer science or CT into their curricula and learning frameworks for early levels of education. These can be helpful for increasing access to quality CT education. However, the use of digital technology in ECEC should add to children's experiences rather than replace interactions with traditional learning materials and games*". Many European Union (EU) Member States have recently revised their curricula by introducing basic CS concepts into compulsory education for developing students' CT skills as a part of the Digital Education Action Plan 2021–2027, where computing education is one of the prerequisites outlined by the priority "Enhancing digital skills and competences for the digital transformation" [125]. Curriculum solutions combining gamification and unplugged and low-cost design elements can better train children's CT in kindergarten [108].

To bridge the gap between integrating CT into ECE curricula and its practical implementation, the role of educators becomes pivotal. Ensuring that teachers are well-equipped with the necessary skills and knowledge through professional development is essential for effectively applying innovative approaches like unplugged coding and gamified learning in early childhood settings. The need for professional development among early childhood educators is emphasized in several studies and plays a critical role in the successful application of unplugged coding activities and

Table 3 Integration of computational thinking (CT) in early childhood education (ECE)

Key aspect	Findings	Supporting studies
Global recognition	Many countries have incorporated CT into educational frameworks as part of their digital education strategies	[63, 125]
Curriculum integration	EU member states have introduced CT and CS concepts into compulsory education through the digital education action plan	[125]
Challenges	Ensuring CT education remains interactive and enhances traditional learning rather than replacing it	[63]
Long-term research	Ongoing research is needed to refine curricula and evaluate the impact of CT integration	[2, 126]

gamified learning approaches in ECE [108]. Saxena et al. [36] also highlight the need to provide professional training especially for those teachers with no or limited CT knowledge as adequate training ensures that educators can effectively integrate coding and programming activities into their teaching, thereby enhancing children's digital competences [127, 128]. Akiba [129] argues that "*Currently, most of the efforts in standard early childhood settings, taught by teachers who are not digital experts, appear to revolve around "unplugged" activities that do not directly involve digital tools or devices, and it is not entirely clear how well these "unplugged" lessons promote the corresponding skills in "plugged" settings, such as coding and programming*". Pila et al. [130] state that professional programs and training have to meet the educators' needs while supporting them to understand how to use technology in an effective way. Table 4 collectively presents analysis outcomes regarding the pedagogical strategies.

Ongoing research is essential to understand the long-term impacts of CT in ECE (Table 5). Studies are needed to evaluate effective teaching methods and help refine curricula and teaching strategies, ensuring they meet the evolving needs of both students and educators [2, 126]. The integration of robotics, unplugged activities and board games in ECE appears to be an innovative method for developing CT. These tools have motivated students in ECE to engage with the core CT components and enhance their problem-solving, and creative expression through interactive and hands-on experiences [79, 81, 131, 132, 134]. Furthermore, research reinforces their wider impact and differentiated learning, underlining their potential for interdisciplinary learning [69, 135]. All these tools, activities and strategies must be tailored to students' developmental stage, emphasizing play and exploration [136]. Furthermore, it is very important to ensure that CT education is inclusive, providing all children—regardless of their background or their ability—with access to these skills [63] while ensuring equitable, effective, and sustainable integration of CT in ECE [57].

Table 4 Effective pedagogical strategies for CT in ECE

Key Aspect	Findings	Supporting studies
Educational robotics (bee-bot, blue-bot, KIBO, etc.)	Enhances engagement, problem-solving, and hands-on learning	[79, 81, 131]
Unplugged activities (board games, storytelling, hands-on play)	Supports pattern recognition, algorithmic thinking, and collaboration	[108, 129]
Gamified learning	Increases motivation and interactive engagement with CT concepts	[108, 132]
Teacher training and professional development	Essential for effective CT integration; helps educators implement innovative methods	[36, 127, 130]

Table 5 Challenges and future directions in CT integration

Challenge	Proposed solutions	Supporting studies
Limited teacher training	Professional development programs to equip educators with CT knowledge	[36, 128]
Equity and inclusion	Ensuring all children, regardless of background, have access to CT education	[57, 63]
Measuring learning outcomes	More research is needed on the impact of unplugged vs. plugged activities	[129]
Future research directions	Need for long-term studies on CT's impact on early learners	[24, 133]

Anticipating the future, the role of CT in ECE looks promising, as educators and policymakers increasingly recognize its importance [63, 133, 137]. CT tools and technologies provide new opportunities for interactive learning that can further enhance children's engagement with CT concepts, by establishing an environment that nurtures creativity and curiosity through age-appropriate practices, investing in educators training and effective strategies [19, 24]. In conclusion, the integration of CT into early education is no longer optional but essential in today's digital age [138–140]. It is vital to empower students with the essential skills for a future where technology is deeply embedded in all aspects of life [63, 141]. A commitment to equip future generations to become active creators of technology and drive innovation in an ever-evolving technological world [23, 62].

References

1. Voogt, J., Fisser, P., Good, J., Mishra, P., Yadav, A.: Computational thinking in compulsory education: towards an agenda for research and practice. Educ. Inf. Technol. **20**, 715–728 (2015)
2. Bocconi, S., Chioccariello, A., Dettori, G., Ferrari, A., Engelhardt, K.: Developing Computational Thinking in Compulsory Education—Implications for Policy and Practice (EUR 28295 EN). European Commission, Brussel (2016). https://doi.org/10.2791/792158
3. Wing, J.M.: Computational thinking. Commun. ACM. **49**(3), 33–35 (2006). https://doi.org/10.1145/1118178.1118215
4. Papert, S.: Mindstorms: Children, Computers, and Powerful Ideas. Basic Books, New York (1980)
5. CSTA, ISTE.: Computational Thinking in K-12 Education Leadership Toolkit (2011)
6. Brennan, K., & Resnick, M.: New frameworks for studying and assessing the development of computational thinking. In: Proceedings of the 2012 Annual Meeting of the American Educational Research Association, pp. 1–25 (2012)
7. Csizmadia, A.P., Curzon, P., Dorling, M., Humphreys, S., Ng, T., Selby, C.C., Woollard, J.: Computational Thinking: A Guide for Teachers. Computing at School, Swindon (2015)
8. Grover, S., Pea, R.: Computational thinking in K–12: a review of the state of the field. Educ. Res. **42**(1), 38–43 (2013). https://doi.org/10.3102/0013189X12463051
9. Susanti, R.D., Lukito, A., Ekawati, R.: Pattern recognition in computational thinking: semiotic perspective. Math. Teach. Res. J. **16**(1), 214 (2024)

10. Rosali, D.F., Suryadi, D.: An analysis of students' computational thinking skills on the number patterns lesson during the Covid-19 pandemic. Formative Sci. J. MIPA Educ. **11**(2), 217–232 (2021). https://doi.org/10.30998/formatif.v11i2.9905
11. Wang, D., Wang, T., Liu, Z.: A tangible programming tool for children to cultivate computational thinking. Sci. World J. **2014**, 428080 (2014). https://doi.org/10.1155/2014/428080
12. Barr, V., Stephenson, C.: Bringing computational thinking to K-12: what is involved and what is the role of the computer science education community? ACM Inroads. **2**(1), 48–54 (2011). https://doi.org/10.1145/1929887.1929905
13. Kalelioğlu, F., Gülbahar, Y., Kukul, V.: A framework for computational thinking based on a systematic research review. Balt. J. Mod. Comput. **4**(3), 583–596 (2016). https://www.bjmc.lu.lv/fileadmin/user_upload/lu_portal/projekti/bjmc/Contents/4_3_15_Kalelioglu.pdf
14. Shute, V.J., Sun, C., Asbell-Clarke, J.: Demystifying computational thinking. Educ. Res. Rev. **22**, 142–158 (2017). https://doi.org/10.1016/j.edurev.2017.09.003
15. Wing, J.M.: Research Notebook: Computational Thinking—What and Why? The Link Magazine, Spring. Carnegie Mellon University, Pittsburgh (2011)
16. National Research Council (NRC): Report of a Workshop on the Scope and Nature of Computational Thinking. The National Academies Press, Washington (2010). https://doi.org/10.17226/12840
17. Wing, J.M.: Computational thinking's influence on research and education for all. Ital. J. Educ. Technol. **25**(2), 7–14 (2017). https://doi.org/10.17471/2499-4324/922
18. Rich, P., Langton, M.: Computational thinking: toward a unifying definition. In: Spector, M., Lockee, B., Childress, M. (eds.) Learning, Design, and Technology: an International Compendium of Theory, Research, Practice, and Policy, pp. 419–429. Springer, Heidelberg (2016). https://doi.org/10.1007/978-3-319-30295-9_14
19. Yadav, A., Mayfield, C., Zhou, N., Hambrusch, S., Korb, J.T.: Computational thinking in elementary and secondary teacher education. ACM Trans. Comput. Educ. **14**(1), 5 (2014). https://doi.org/10.1145/2576872
20. Lee, I., et al.: Computational thinking for youth in practice. ACM Inroads. **2**(1), 32–37 (2011). https://doi.org/10.1145/1929887.1929902
21. Snyder, H.: Literature review as a research methodology: an overview and guidelines. J. Bus. Res. **104**, 333–339 (2019)
22. Bers, M.: From computational thinking to computational doing. In: Research Anthology on Computational Thinking, Programming, and Robotics in the Classroom, pp. 1–15. IGI Global, Hershey (2021). https://doi.org/10.4018/978-1-7998-7308-2.ch001
23. Wyeth, P., Wyeth, G.: Robot building for preschoolers. In: RoboCup 2007: Robot Soccer World Cup XI, LNCS, vol. 5001, pp. 124–135. Springer, Heidelberg (2008)
24. Mezak, J., Pejić Papak, P., Vujičić, L.: The integration of algorithmic thinking into preschool education. In: EDULEARN21 Proceedings, pp. 8182–8187 (2021)
25. Wong, G.K.W., Jian, S., Cheung, H.Y.: Engaging children in developing algorithmic thinking and debugging skills in primary schools: a mixed-methods multiple case study. Educ. Inf. Technol. **29**(1), 16205–16254 (2024). https://doi.org/10.1007/s10639-024-12448-x
26. Masarwa, B., Hel-Or, H., Levy, S.T.: Kindergarten children's learning of computational thinking with the "sorting like a computer" learning unit. J. Res. Child. Educ. **38**(2), 165–188 (2023). https://doi.org/10.1080/02568543.2023.2221319
27. Vujičić, L., Jančec, L., Mezak, J.: Development of algorithmic thinking skills in early and preschool education. In: EDULEARN21 Proceedings, pp. 8152–8161. IATED, Valencia (2021)
28. Acosta, Y., Alsina, Á., Pincheira, N.: Computational thinking and repetition patterns in early childhood education: longitudinal analysis of representation and justification. Educ. Inf. Technol. **29**(5), 7633–7658 (2024). https://doi.org/10.1007/s10639-023-12051-6
29. Piaget, J.: The Grasp of Consciousness: Action and Concept in the Young Child. Harvard University Press, Cambridge (1976)

30. Yuriansa, A., Kurniawati, L.: The importance of playing pattern for early childhood mathematics learning. In: Proceedings of the 3rd International Conference on Learning Innovation and Quality Education (ICLIQE 2020), pp. 123–129. Atlantis Press, Dordrecht (2021). https://doi.org/10.2991/assehr.k.210322.019
31. Wu, T.-T., Asmara, A., Huang, Y.-M., Permata Hapsari, I.: Identification of problem-solving techniques in computational thinking studies: systematic literature review. SAGE Open. **14**(2) (2024). https://doi.org/10.1177/21582440241249897
32. Miller, J.: STEM education in the primary years to support mathematical thinking: using coding to identify mathematical structures and patterns. ZDM. **51**(6), 915–927 (2019). https://doi.org/10.1007/s11858-019-01096-y
33. Mulligan, J., Mitchelmore, M.: Awareness of pattern and structure in early mathematical development. Math. Educ. Res. J. **21**(2), 33–49 (2009)
34. Zippert, E.L., Douglas, A.-A., Rittle-Johnson, B.: Finding patterns in objects and numbers: repeating patterning in pre-K predicts kindergarten mathematics knowledge. J. Exp. Child Psychol. **200**, 104965 (2020). https://doi.org/10.1016/j.jecp.2020.104965
35. Calderon, A., Crick, T., Tryfona, C.: Developing computational thinking through pattern recognition in early years education. In: 2015 ACM Conference on Innovation and Technology in Computer Science Education, pp. 259–260. ACM, New York (2015). https://doi.org/10.1145/2783446.2783600
36. Saxena, A., Lo, C.K., Hew, K.F., Wong, G.K.W.: Designing unplugged and plugged activities to cultivate computational thinking: an exploratory study in early childhood education. Asia-Pac. Educ. Res. **29**(1), 55–66 (2020). https://doi.org/10.1007/s40299-019-00478-w
37. Lüken, M.M., Kampmann, R.: The influence of fostering children's patterning abilities on their arithmetic skills in grade 1. In: Elia, I., Mulligan, J., Anderson, A., Baccaglini-Frank, A., Benz, C. (eds.) Contemporary Research and Perspectives on Early Childhood Mathematics Education, pp. 55–66. Springer, Cham (2018). https://doi.org/10.1007/978-3-319-73432-3_4
38. NCTM: Principles and Standards. National Council of Teachers of Mathematics, Reston (2020)
39. Wing, J.M.: Computational thinking and thinking about computing. Philos. Trans. R. Soc. A. **366**, 3717–3725 (2008)
40. Kim, S., Choe, I., Kaufman, J.: The development and evaluation of the effect of creative problem-solving program on young children's creativity and character. Think. Skills Creat. **33**, 100590 (2019). https://doi.org/10.1016/j.tsc.2019.100590
41. Qian, Y., Choi, I.: Tracing the essence: ways to develop abstraction in computational thinking. Educ. Technol. Res. Dev. **3**, 1055–1078 (2022). https://doi.org/10.1007/s11423-022-10182-0
42. Sullivan, A., Bers, M.U.: Robotics in the early childhood classroom: learning outcomes from an 8-week robotics curriculum in pre-kindergarten through second grade. Int. J. Technol. Des. Educ. **26**, 3–20 (2015). https://doi.org/10.1007/s10798-015-9304-5
43. Elkin, M., Sullivan, A., Bers, M.U.: Implementing a robotics curriculum in an early childhood Montessori classroom. J. Inf. Technol. Educ. Innov. Pract. **13**, 153–169 (2014). https://doi.org/10.28945/2094
44. Ezeamuzie, N.O., Leung, J.S.C., Ting, F.S.T.: Unleashing the potential of abstraction from cloud of computational thinking: a systematic review of literature. J. Educ. Comput. Res. **59**(8), 1387–1418 (2021). https://doi.org/10.1177/07356331211055379
45. Saig, R., Hershkovitz, A.: Expanding digital literacies beyond the digital: infusing computational thinking into unplugged pedagogical tools—two case studies from mathematics education. Int. J. Child Comput. Interact. **42**, 100703 (2024). https://doi.org/10.1016/j.ijcci.2024.100703
46. Ho, C.H.: Some phenomena of problem decomposition strategy for design thinking: differences between novices and experts. Des. Stud. **22**(1), 27–45 (2001)
47. Moschella, M.: Observable computational thinking skills in primary school children: How and when teachers can discern abstraction, decomposition and use of algorithms. In: 13th International Technology, Education and Development Conference (INTED), pp. 6259–6267 (2019). https://doi.org/10.21125/inted.2019.1523

48. Rijke, W., Bollen, L., Eysink, T., Tolboom, J.: Computational thinking in primary school: an examination of abstraction and decomposition in different age groups. Inform. Educ. **17**(1), 77–92 (2018). https://doi.org/10.15388/infedu.2018.05
49. Dietz, G., Landay, J.A., Gweon, H.: Building blocks of computational thinking: young children's developing capacities for problem decomposition. In: CogSci, pp. 1647–1653. Cognitive Science Society, Austin (2019)
50. Zeng, Y., Yang, W., Bautista, A.: Teaching programming and computational thinking in early childhood education: a case study of content knowledge and pedagogical knowledge. Front. Psychol. **14** (2023). https://doi.org/10.3389/fpsyg.2023.1252718
51. Ehsan, H., Rehmat, A.P., Cardella, M.E.: Computational thinking embedded in engineering design: capturing computational thinking of children in an informal engineering design activity. Int. J. Technol. Des. Educ. **31**(3), 441–464 (2021). https://doi.org/10.1007/s10798-020-09562-5
52. Angeli, C., Valanides, N.: Developing young children's computational thinking with educational robotics: an interaction effect between gender and scaffolding strategy. Comput. Hum. Behav. **105**, 105954 (2020). https://doi.org/10.1016/j.chb.2019.03.018
53. Zeng, Y., Yang, W., Bautista, A.: Computational thinking in early childhood education: reviewing the literature and redeveloping the three-dimensional framework. Educ. Res. Rev. **39**, 100520 (2023). https://doi.org/10.1016/j.edurev.2023.100520
54. DeLiema, D., Dahn, M., Flood, V.J., Asuncion, A., Abrahamson, D., Enyedy, N., Steen, F.: Debugging as a context for fostering reflection on critical thinking and emotion. In: Deeper Learning, Dialogic Learning, and Critical Thinking, p. 20. Routledge, London (2019). https://doi.org/10.4324/9780429323058
55. Pugnali, A., Sullivan, A., Bers, M.U.: The impact of user interface on young children's computational thinking. J. Inf. Technol. Educ. Innov. Pract. **16**, 171–193 (2017)
56. Sun, C., Yang, S., Becker, B.: Debugging in computational thinking: a meta-analysis on the effects of interventions on debugging skills. J. Educ. Comput. Res. **62**(4), 1087–1121 (2024). https://doi.org/10.1177/07356331241227793
57. Guss, S.S., Clements, D.H., Sharifnia, E., Sarama, J., Holland, A., Lim, C.-I., Vinh, M.: Designing inclusive computational thinking learning trajectories for the youngest learners. Educ. Sci. **14**(7), 733 (2024). https://doi.org/10.3390/educsci14070733
58. Misirli, A., Komis, V.: Computational thinking in early childhood education: the impact of programming a tangible robot on developing debugging knowledge. Early Child Res. Q. **65**, 139–158 (2023). https://doi.org/10.1016/j.ecresq.2023.05.014
59. Ahn, J., Sung, W., Black, J.B.: Unplugged debugging activities for developing young learners' debugging skills. J. Res. Child. Educ. **36**(3), 421–437 (2021). https://doi.org/10.1080/02568543.2021.1981503
60. Bers, M.U., Flannery, L., Kazakoff, E.R., Sullivan, A.: Computational thinking and tinkering: exploration of an early childhood robotics curriculum. Comput. Educ. **72**, 145–157 (2014). https://doi.org/10.1016/j.compedu.2013.10.020
61. Silvis, D., Lee, V., Clarke-Midura, J., Shumway, J.: The technical matters: young children debugging (with) tangible coding toys. Inf. Learn. Sci. **123**, 109 (2022). https://doi.org/10.1108/ILS-12-2021-0109
62. Bers, M.U.: Coding as a Playground: Programming and Computational Thinking in the Early Childhood Classroom, 2nd edn. Routledge, London (2021)
63. Bers, M., Strawhacker, A., Sullivan, A.: The state of the field of computational thinking in early childhood education. In: OECD Education Working Papers, No. 274. OECD Publishing, Paris (2022). https://doi.org/10.1787/3354387a-en
64. Silvis, D., Lee, V., Clarke-Midura, J., Shumway, J., Kozlowski, J.: Blending everyday movement and representational infrastructure: an interaction analysis of kindergarteners coding robot routes. In: Gresalfi, M., Horn, I.S. (eds.) ICLS 2020, vol. 1, pp. 98–105. ISLS, Helsinki (2020)
65. Druin, A., Hendler, J.: Robots for Kids: Exploring New Technologies for Learning. Academic Press, San Diego (2000)

66. Busuttil, L., Vassallo, D., Callus, J., Bratitsis, T., Tsapara, M., Melliou, K., Meireles, G., Tarraf Kojok, N., Koliakou, I., & Sousa, S.: Promoting computational thinking in early childhood: A novel board game design. In: ICERI2023 Proceedings, pp. 4868–4876 (2023)
67. Zviel-Girshin, R., Rosenberg, N., Kukliansky, I.: Early childhood robotics: children's beliefs and objective capabilities to read and write programs. J. Res. Child. Educ. **38**(2), 317–335 (2023). https://doi.org/10.1080/02568543.2023.2259946
68. Bers, M.: Blocks to Robots: Learning with Technology in the Early Childhood Classroom. Teacher's College Press, New York (2008)
69. Sullivan, A., Kazakoff, E.R., Bers, M.U.: The wheels on the bot go round and round: robotics curriculum in pre-kindergarten. J. Inf. Technol. Educ. Innov. Pract. **12**, 203 (2013)
70. Sinha, R., Swanson, H., Clarke-Midura, J., Shumway, J.F., Lee, V.R., Chandrasekharan, S.: From embodied doing to computational thinking in kindergarten: A punctuated motor control model. In: Proc. of the 2023 Symp. on Learning, Design and Technology, pp. 1–10 (2023)
71. Zviel-Girshin, R., Luria, A., Shaham, C.: Robotics as a tool to enhance technological thinking in early childhood. J. Sci. Educ. Technol. **29**(2), 294–302 (2020). https://doi.org/10.1007/s10956-020-09815-x
72. Flannery, L.P., Bers, M.U.: Let's dance the "robot hokey-pokey!" children's programming approaches and achievement throughout early cognitive development. J. Res. Technol. Educ. **46**(1), 81–101 (2013)
73. Metin, S.: Activity-based unplugged coding during the preschool period. Int. J. Technol. Des. Educ. **32**, 149–165 (2022). https://doi.org/10.1007/s10798-020-09616-8
74. Chaldi, D., Mantzanidou, G.: Educational robotics and STEAM in early childhood education. Adv. Mobile Learn. Educ. Res. **1**(2), 72–81 (2021)
75. João-Monteiro, M., Cristóvão-Morgado, R., Bulas-Cruz, M., Morgado, L.: A robot in kindergarten. In: Eurologo'2003 Proceedings - Re-inventing Technology on Education, pp. 1–6 (2003)
76. Kilianova, K., Kockova, P., Bohackova, P., Klubal, L., Kostolanyova, K.: Bee-bot & Blue-bot: from mindless tapping to algorithm creation. In: ICERI2022 Proceedings, pp. 5565–5572. IATED, Valencia (2022)
77. Newhouse, C.P., Cooper, M., Cordery, Z.: Programmable toys and free play in early childhood classrooms. Aust. Educ. Comput. **32**(1), 147 (2017)
78. Christidi, Z., Christopoulou, E.: Environmental awareness in preschool education via educational robotics and STEAM education. In: 7th South-East Europe Design Automation, Computer Engineering, Computer Networks and Social Media Conference (SEEDA-CECNSM), pp. 1–8. IEEE, New York (2022). https://doi.org/10.1109/SEEDA-CECNSM57760.2022.9932911
79. Alvarez, J., Bellegarde, K., Flahaut, J.-J., Lafouge, T.: Blue bot project experiment: first statistical results of the performance of 5-year-old pupils on their ability to program. In: 8th International Toy Research Association (ITRA) World Conference. International Toy Research Association, Rochester (2018)
80. Ferrara, F., Ferrari, G., Savioli, K.: Children in movement towards STEAM: coding and shapes at kindergarten. In: 15th International Conference on Technology in Mathematics Teaching (ICTMT 15), pp. 1–6. University of Turin, Turin (2022). https://iris.unito.it/handle/2318/1886208
81. Gambin, S.E.: The Bee-Bot in the Assessment for Learning of Mathematical Concepts. Univ. Malta Institutional Repository, Malta (2023). https://www.um.edu.mt/library/oar/handle/123456789/118574
82. Pinto, M., Fernandes, M., Osório, A.J.: Curriculum integration proposal for KIBO robot in preschool: Kids media lab project. In: INTED2021 Proceedings, pp. 7910–7921 (2021)
83. Elkin, M., Sullivan, A., Bers, M.U.: Programming with the KIBO robotics kit in preschool classrooms. Comput. Sch. **33**(3), 169–186 (2016)
84. Miková, K., Fejková, V.: A comparison of two (not well known) educational robotic kits for teaching computational thinking. In: ICERI2022 Proceedings, pp. 7830–7837. IATED, Valencia (2022)

85. Zurnacı, B., Turan, Z.: Educational robotics or unplugged coding activities in kindergartens?: comparison of the effects on pre-school children's computational thinking and executive function skills. Think. Skills Creat. **53**, 101576 (2024). https://doi.org/10.1016/j.tsc.2024.101576
86. Di Donato, D., Mattioli, P.: Beyond barriers. Inclusion and innovation through the use of educational robotic environments. In: Zecca, L., Datteri, E. (eds.) Inclusive Science Education and Robotic, pp. 159–172. FrancoAngeli, Milano (2022)
87. Montuori, C., et al.: Combined unplugged and educational robotics training to promote computational thinking and cognitive abilities in preschoolers. Educ. Sci. **13**(9), 858 (2023). https://doi.org/10.3390/educsci13090858
88. Berland, M., Lee, V.R.: Collaborative strategic board games as a site for distributed computational thinking. Int. J. Game-Based Learn. **1**(2), 65 (2011). https://doi.org/10.4018/ijgbl.2011040104
89. Chen, S., Tsai, J., Liu, S., Chang, C.: The effect of a scientific board game on improving creative problem-solving skills. Think. Skills Creat. **41**, 100894 (2021). https://doi.org/10.1016/j.tsc.2021.100894
90. Romero, M., Viéville, T., Duflot-Kremer, M.: Activity for Learning Computational Thinking in Plugged and Unplugged Mode (Research Report No. 006). UCA-INSPE Académie de Nice, Nice (2022). https://hal.archives-ouvertes.fr/hal-03793719
91. Lin, Y., et al.: Comparing the effects of plugged-in and unplugged activities on computational thinking development in young children. Educ. Inf. Technol. **29**, 9541–9574 (2024). https://doi.org/10.1007/s10639-023-12181-x
92. Bati, K.: A systematic literature review regarding computational thinking and programming in early childhood education. Educ. Inf. Technol. **27**, 2059–2082 (2022). https://doi.org/10.1007/s10639-021-10700-2
93. Brennan, K., & Resnick, M.: Using artifact-based interviews to study the development of computational thinking in interactive media design. In: Annual American Educational Research Association Meeting. Vancouver, BC, Canada (2012)
94. Robinson, A.: Unplugged Learning in the Kindergarten Computer Science Classroom. Northwestern College - Orange City (2022)
95. Chen, P., Yang, D., Metwally, A.H.S., et al.: Fostering computational thinking through unplugged activities: a systematic literature review and meta-analysis. Int. J. STEM Educ. **10**(47) (2023). https://doi.org/10.1186/s40594-023-00434-7
96. Küçükkara, M.F., Aksüt, P.: An example of unplugged coding education in preschool period: activity based algorithm for problem solving skills. J. Inq. Based Act. **11**(2), 81–91 (2021). https://www.ated.info.tr/ojs-3.2.1-3/index.php/ated/issue/view/22
97. Mladenović, M., Repušić, M., Ozretić, M.: CS unplugged activities for teaching basic programming concepts to kindergarten children. In: ICERI 2023 Proceedings (2023). https://doi.org/10.21125/iceri.2023.1926
98. Cakiroglu, U., Mumcu, S., Atabayf, M., Aydin, M.: Understanding problem-solving processes of preschool children in CS unplugged activities. Int. J. Comput. Sci. Educ. Sch. **5**(3), 35–53 (2022). https://doi.org/10.21585/ijcses.v5i3.133
99. Lee, J.: Coding in early childhood. Contemp. Issues Early Child. **21**(3), 266–269 (2020). https://doi.org/10.1177/1463949119846541
100. Lee, J., Junoh, J.: Implementing unplugged coding activities in early childhood classrooms. Early Child. Educ. J. **47**, 709–716 (2019). https://doi.org/10.1007/s10643-019-00967-z
101. Bratitsis, T., Busuttil, L., Vassallo, D., Koliakou, I., Tsapara, M., Melliou, K., Callus, J., Meireles, G., Tarraf Kojok, N., Sousa, S., Arnaoutis, G.: Cthink. It. A board-game for computational thinking in early years. In: 18th European Conference on Games Based Learning. Academic Conferences International Limited, Reading (2024)
102. Tsapara, M., Melliou, K., Bratitsis, T.: Cultivating computational thinking through board games in early childhood. In: 11th Panhellenic Conf. on Informatics Education. https://eproceedings.epublishing.ekt.gr/index.php/cetpe/article/view/7221 (2023)

103. Daily, S.B., Leonard, A.E., Jörg, S., Babu, S., Gundersen, K.: Dancing Alice: exploring embodied pedagogical strategies for learning computational thinking. In: In: 45th ACM Technical Symposium on Computer Science Education, pp. 91–96. ACM, New York (2014)
104. Budinski, N., Lavicza, Z., Fenyvesi, K., Novta, M.: Mathematical and coding lessons based on creative origami activities. Open Educ. Stud. **1**(1), 220–227 (2019)
105. Tsarava, K., Moeller, K., Pinkwart, N., Butz, M., Trautwein, U., Ninaus, M.: Training computational thinking: Game-based unplugged and plugged-in activities in primary school. In: 11th Eur. Conf. Games Based Learn., pp. 687–695 (2017)
106. Rehder, M.M., Jensen, J.J., Møller, T.E., Schrøder, V.: Exploring the potentials of unplugged activities: developing computational thinking in teacher education. Nord. J. Comp. Int. Educ. **8**(4), 5741 (2024)
107. Zhang, Y., Liang, Y., Tian, X., Yu, X.: The effects of unplugged programming activities on K-9 students' computational thinking: meta-analysis. Educ. Technol. Res. Dev. **72**, 1331–1356 (2024)
108. Özdinç, F., Kandemir-Özdinç, N.: Using interdisciplinary unplugged CS activities to support the development of preschool children. In: INTED2024 Proceedings, pp. 7492–7496 (2024)
109. Wang, L., Huang, M., Ismail Kamal, J.B.: The application of an unplugged and low-cost children's coding education tool in a gamification context. Pertanika J. Sci. Technol. **32**(2) (2024). https://doi.org/10.47836/pjssh.32.2.03
110. Bayeck, R.Y.: Examining board gameplay and learning: a multidisciplinary review of recent research. Simul. Gaming. **51**(4), 411–431 (2020). https://doi.org/10.1177/1046878119901286
111. Ching, Y.H., Hsu, Y.C., Baldwin, S.: Developing computational thinking with educational technologies for young learners. TechTrends. **62**(6), 563–573 (2018). https://doi.org/10.1007/s11528-018-0292-7
112. Apostolellis, P., Stewart, M., Frisina, C., Kafura, D.: RaBit EscAPE: A board game for computational thinking. In: Proceedings of the 2014 Conference on Interaction Design and Children, pp. 349–352. ACM, New York (2014). https://doi.org/10.1145/2593968.2610485
113. Tsarava, K., Moeller, K., Ninaus, M.: Board games for training computational thinking. Lect. Notes Comput. Sci. **9**, 11385 (2019). https://doi.org/10.1007/978-3-030-11548-7_9
114. Eisenack, K.: A climate change board game for interdisciplinary communication and education. Simul. Gaming. **43**(2), 1–21 (2012). https://doi.org/10.1177/1046878112452639
115. Zirawaga, S., Adeleye, O., Tinovimbanashe, M.: Gaming in education: using games as a support tool to teach history. J. Educ. Pract. **8**(15), 55–64 (2017)
116. Tsarava, K., Moeller, K., Ninaus, M.: Training computational thinking through board games: the case of Crabs & Turtles. Int. J. Serious Games. **5**(2), 25–44 (2018). https://doi.org/10.17083/ijsg.v5i2.248
117. Scirea, M., Valente, A.: Board games and computational thinking: how to identify games with potential to support CT in the classroom. In: 14th EAI International Conference on Pervasive Computing Technologies for Healthcare, pp. 349–358. ACM, New York (2020). https://doi.org/10.1145/3402942.3409616
118. Erdogan, A., Atasay Sunay, M., Eryılmaz Çevirgen, A.: How to play board games? A framework proposal for classroom settings. Board Game Stud. J. **16**(2), 101–122 (2022). https://doi.org/10.2478/bgs-2022-0021
119. Eicher, L.: The impact of scaffolding through the use of board games on student goal orientation. Honors Projects, 118. Bowling Green State University (2014). https://scholarworks.bgsu.edu/honorsprojects/118
120. Hunsucker, A.J.: Board games as a platform for collaborative learning. In: Meaningful Play 2016, pp. 1–10 (2016)
121. Uribe, V.R., Cobos, S.: Board-games as a tool for teaching basic sustainability concepts to design students. In: Eur. Conf. on Sustainability, Energy and the Environment. IAFOR, Toyonaka (2014)
122. Tsai, J.C., Liu, S.Y., Chang, C.Y., Chen, C.Y.: Using a board-game to teach about sustainable development. Sustainability. **13**(9), 4942 (2021)

123. Machuqueiro, F., Piedade, J.: Game on: a journey into computational thinking with modern board games in Portuguese primary education. Educ. Sci. **14**(11), 1182 (2024). https://doi.org/10.3390/educsci14111182
124. Tsapara, M., Bratitsis, T.: Board-game design by children as an assessment mechanism in kindergarten. A case study about disability and vulnerability. In: Smart Mobile Communication & Artificial Intelligence, pp. 90–105. Springer, Cham (2024)
125. Bocconi, S., Chioccariello, A., Kampylis, P., Dagienė, V., Wastiau, P., Engelhardt, K., Stupurienė, G.: Reviewing Computational Thinking in Compulsory Education: State of Play and Practices from Computing Education (2022)
126. Ngadengon, A., Zuraihah, Z., Subramaniam, T., Yasak, Z., Syukri, M., Hazim, N.: Theory on computational thinking in education: a systematic review. Int. J. Educ. Psychol. Couns. **9**, 488–507 (2024). https://doi.org/10.35631/IJEPC.953037
127. Rajapakse Mohottige, N.U.S., Bjerke, A.H., Andersen, R.: Teacher education as stakeholder: teacher educator perspectives on the integration of computational thinking into mathematics and science courses. Teachers Teach. **30**(4), 437–451 (2024). https://doi.org/10.1080/13540602.2024.2313635
128. Kravik, R., Berg, T.K., Siddiq, F.: Teachers' understanding of programming and computational thinking in primary education–a critical need for professional development. Acta Didact. Norden. **16**(4), 23 (2022)
129. Akiba, D.: Computational thinking and coding for young children: a hybrid approach to link unplugged and plugged activities. Educ. Sci. **12**(11), 793 (2022). https://doi.org/10.3390/educsci12110793
130. Pila, S., Blackwell, C.K., Lauricella, A.R., Wartella, E.: Technology in the Lives of Educators and Early Childhood Programs. Northwestern University, Evanston (2013)
131. Korenova, L., Gasparova, E., Kostrub, D.: Interpretation of causal relationships in the world of robots presented by 5-year-old children. In: INTED2022 Proceedings, pp. 300–306. IATED, Valencia (2022)
132. Di Lieto, M.C., Pecini, C., Castro, E., Inguaggiato, E., Cecchi, F., Dario, P., Sgandurra, G.: Empowering executive functions in 5- and 6-year-old typically developing children through educational robotics: an RCT study. Front. Psychol. **10**, 3084 (2020). https://doi.org/10.3389/fpsyg.2020.03084
133. Yang, W., Su, J., Li, H.: Demystifying early childhood computational thinking: an umbrella review to upgrade the field. Future Educ. Res. **2**(4), 458–477 (2024). https://doi.org/10.1002/fer3.38
134. Samuelsson, R.: Cultural heritage through educational robots: using a Ukrainian folk tale with a programmable robot in early childhood education. In: HCII 2022—Late Breaking Posters, LNCS, vol. 13338, pp. 336–343. Springer, Heidelberg (2022)
135. Liu, E.Z.F., Lin, C.H., Liou, P.Y., Feng, H.C., Hou, H.T.: An analysis of teacher-student interaction patterns in a robotics course for kindergarten children: a pilot study. Turk. Online J. Educ. Technol. **12**(1), 9–18 (2013)
136. Lavigne, H., Orr, J., Wolsky, M.: Message in a backpack: helping your preschool child with computational thinking. Teach. Young Child. **15**, 3 (2022)
137. González-Sancho, C.: Computational Thinking for Young Kids. OECD, Paris (n.d.)
138. Mestic, G., et al.: Implications of computational thinking in the initial training of future primary and pre-school teachers. Educatia. **21**, 21–28 (2024). https://doi.org/10.24193/ed21.2024.28.21
139. Alonso-García, S., Fuentes, A., Navas-Parejo, M., Maldonado, J.V.: Enhancing computational thinking in early childhood education with educational robotics: a meta-analysis. Heliyon. **10**, e33249 (2024). https://doi.org/10.1016/j.heliyon.2024.e33249
140. Isikoglu, N.: New toys for young children: integration of computer technology into early childhood education. Turk. Online J. Educ. Technol. **2**(4), 5 (2003)
141. Saidin, N.D., Khalid, F., Martin, R., Kuppusamy, Y., Munusamy, N.: Benefits and challenges of applying computational thinking in education. Int. J. Inf. Educ. Technol. **11**(5), 248–254 (2021). https://doi.org/10.18178/ijiet.2021.11.5.1519

Teaching Fractions to Third-Grade Students Using Enriched Books: A Case Study

Vasiliki Georgoulia, Aggeliki Tzavara, Andromachi Filippidi, Konstantinos Lavidas, and Vassilis Komis

Abstract This study examines the outcomes of a teaching intervention aimed at enhancing third-grade students' understanding of fractions in Greece through the use of enriched books from the Digital School. Aligned with the third-grade mathematics curriculum, the intervention included initial and final assessments, along with four intermediate tests during the teaching process. Data were analyzed using NVivo software, enabling a thematic analysis of (a) the effectiveness of integrating traditional teaching methods with digital tools in fraction learning, (b) student engagement, and (c) the teacher's role in the process. The findings indicate that using enriched books positively impacts student performance and fosters a deeper understanding of new concepts. Moreover, the study highlights the teacher's multifaceted role in providing essential support to students throughout the learning process.

Keywords Enriched books · Digital tools · Teaching intervention · Fractions

1 Introduction

Learning fractions in primary education is a critical yet demanding process that requires specialized teaching approaches and tools. Fractions are a fundamental mathematical concept introduced from the first grades of primary school and develop

V. Georgoulia (✉) · A. Tzavara · A. Filippidi · K. Lavidas · V. Komis
University of Patras, Patra, Greece
e-mail: vgeorgou@hotmail.com

A. Tzavara
e-mail: tzavara@upatras.gr

A. Filippidi
e-mail: afilippidi@upatras.gr

K. Lavidas
e-mail: lavidas@upatras.gr

V. Komis
e-mail: komis@upatras.gr

I. Kazanidis and A. Tsinakos (eds.), *Research on E-Learning and ICT in Education*,
https://doi.org/10.1007/978-3-031-99961-1_2

until the end of primary education. A correct understanding of this concept is important not only for the future learning path of students but also for their everyday lives. Existing research indicates that students often struggle with fractions due to the complexity of the concepts and potential misconceptions [1–3]. In addition, understanding fractions is often accompanied by errors and misconceptions, such as confusion between the numerator and denominator, difficulty in representing fractions on the number line, and generalizing the properties of natural numbers to fractions. These errors can negatively affect students' learning if not addressed early [4–6].

As demonstrated by extensive research [7–9], using information and communication technologies (ICT) in education is particularly effective in enhancing understanding of mathematical concepts. Digital tools, including interactive software and educational games [10], enhance students' comprehension of fraction concepts by offering visual representations and engaging activities. In addition, the role played by the teacher during the learning process is critical, as demonstrated by numerous studies. According to Rindu and Ariyanti [11], a teacher assumes at least eight different roles, the role of a regulator, the role of an evaluator, the role of an organizer/manager, the role of a prompter/motivator, the role of a participant, the role of a source of knowledge, the role of a facilitator, and the role of a guide. These roles support students in acquiring new knowledge and developing metacognitive skills, an essential component of mathematics instruction [12]. In addition, providing clear and explicit guidelines to students for the task they are required to perform is considered particularly important [13–15].

This paper analyzes a teaching intervention designed to enhance third-grade students' understanding of fractions using enriched books from the Digital School. Traditional teaching methods often fail to address the needs and challenges students face when learning fractions. The integration of digital tools into the teaching process is proposed as a modern and effective method to improve the comprehension and reinforcement of mathematical concepts.

Based on the above, this study aims to:

- Evaluate the effectiveness of combining traditional teaching methods with digital tools for learning fractions.
- Examine student performance and conceptual understanding before and after the intervention.
- Investigate the roles of students and teachers during the intervention.

The teaching intervention was based on the recommendations of the third-grade mathematics curriculum and included initial and final assessment tests and four interim tests during the lessons. Thematic analysis of the data was carried out using NVivo software, providing valuable information about the educational process and its results.

Upon completion of the study, the findings are expected to contribute significantly to the understanding of the effectiveness of teaching interventions using digital tools and, through this understanding, to guide the formulation of improved practices and strategies for teaching fractions in primary education.

2 Scope and Description of the Research

2.1 Research Context

For the present study, the teacher/researcher conducted four 2-h teaching interventions with third-grade students for teaching the fourth unit of the third-grade mathematics textbook entitled "Introduction to simple fractions". In this unit, students are taught, for the first time, about fractions, simple fractions, and equivalent fractions. The teacher/researcher maintained an active and consistent presence during learning, particularly when using digital tools. In particular, the thematic teaching approach was used but combined in several cases with the procedural approach [16] using digital tools, while at the same time, in all four teaching interventions, the student's enriched textbook was used, and the micro-experiments suggested by the enriched textbook were used.

In this context, it was investigated:

1. Whether there has been a change in student achievement, using a diagnostic test before and after the end of the teaching interventions.
2. The student's learning path and the gradual change in their performance, both through the activities and through the assessment test of each teaching intervention separately.
3. The role of the teacher in the learning process.

The sample of the study consisted of third-grade students from a Greek primary school. Students participated in four 2-h teaching interventions designed to enhance their understanding of fractions.

2.1.1 The Micro-Experiments

Going into more detail on the micro-experiments by chapter:

- The micro-experiment included in chapter 22, "Introduction to fractions" (https://photodentro.edu.gr/v/item/ds/8521/4180), is a geometric exploration of the relationship between fractions and everything and is called "Simple Fractions". Students, using the click-and-drag facility provided by the micro-experiment, are asked to compare two rectangles and investigate the relationship between them.
- The micro-experiment included in chapter 23, "The fractional units" (https://photodentro.edu.gr/v/item/ds/8521/4172), is a geometric investigation of the relationship between the fractional units and the whole, using the pie model called "Playing with the circle". Students are guided through the micro-experiment and are asked at each division of the circle to identify which fraction is represented.

Chapter 24, "Fractions and Simple Fractions", includes two micro-experiments:

- The first micro-experiment (https://photodentro.edu.gr/v/item/ds/8521/4175) concerns the geometric exploration and interpretation of the relationship between the part and the whole through a realistic activity. The game is called "Fractions for the Sun". Students are asked to answer what part of the window the curtain must cover to hide the sun and express that part of the window with a simple fraction.

- The second micro-experiment concerns the geometric exploration and interpretation of the relationship between the part and the whole, comparing different lengths of scratched pencils. The game is called "Comparing scratched pencils" and we must compare scratched pencils A and B with the original pencil as seen on the screen (https://photodentro.edu.gr/v/item/ds/8521/4173). The task is for the students to find out what part of the pencil is left each time in the form of a fraction.

Chapter 25, "Equivalent Fractions", includes three micro-experiments of which we have used only two:

- The first micro-experiment (https://photodentro.edu.gr/v/item/ds/8521/4167) involves exploring and interpreting equivalent fractions through a realistic activity. The game is called "Who likes chocolate best?". By observing the corresponding part of the chocolate eaten by each of the two children in the experiment being colored, students must answer which child ate more and explain why. Two more similar questions follow.
- The second micro-experiment involves geometric exploration and deepening the understanding of equivalent fractions through a realistic activity. The game is called "Equivalent Fractions - Partner with a Chocolate" (https://photodentro.edu.gr/v/item/ds/8521/4170). Students are asked to solve problems like those in the previous micro-experiment using the chocolate that appears on the screen.

2.1.2 Research Data Collection

To collect the research data, (a) the initial and final assessment test of students' performance that included exercises around fractions (diagnostic test) was created. The diagnostic test was completed once by the students before the start of the teaching interventions and once after the completion of all teaching interventions, and (b) four different assessment tests related to the fractions part of the curriculum of the respective teaching intervention. The assessment tests were answered immediately after the end of each instructional intervention, and they contained six activities:

- The first activity consisted of 3 columns. In the first column, there are 4 pizzas divided into equal parts, with one part colored. The second column gives the algebraic representation of the colored part of the pizza, and the third column gives the verbal representation. The students must match each circle with one of the fractions in the second column and with the words expressing the fractions in the third column.
- In the second activity, we have 3 pizzas divided into equal parts/slices and next to each pizza is written the algebraic representation of a fraction. Students must color the pieces of pizza that represent the given fraction.

- In the third activity, we have a rectangle representing a chocolate bar divided into 12 equal parts. Three children eat a different part of the chocolate given in fractional form. Students are asked to color the part of the chocolate each child ate a different color and then place these three fractions from the smallest to the largest.
- The fourth activity involves 3 circles representing pizzas, divided into equal pieces and next to each circle is a fraction written. Students must color the pieces indicated by the fraction.
- The fifth activity involves three rectangles representing chocolates divided into equal parts and coloring a portion of each rectangle. Below each rectangle is a fraction. Students must match each rectangle with the fraction showing the part of the rectangle that is colored.
- In the sixth activity we have a line segment divided into 9 equal parts. The bicycle has traveled a certain part of the segment. Students have to circle the 2 fractions out of the 5 given to them and express the part of the straight line segment that the bicycle has traveled.

At the same time, we utilized Free Cam 8, which is a desktop screen recording software that allows users to record and share their computer screen activities, either partially or the entire screen and at the same time enables the recording of conversations.

Finally, NVivo, which is a qualitative data analysis software, was utilized for data coding and analysis. The software was fed by the researcher/trainer with the transcripts of the recordings recorded during the teaching interventions.

2.2 Research Questions

In this study, we investigated the effectiveness of a teaching intervention for the teaching of fractions to Greek third-grade students. This instruction, as proposed by the third-grade mathematics curriculum and supported by the enriched textbook, is based on the traditional teaching method with the simultaneous use of digital tools.

The research questions guiding this study are:

- Can the combination of traditional teaching and enriched books improve third-grade students' understanding of fractions?
- How does student participation evolve throughout the teaching intervention?
- What specific instructional strategies enhance learning outcomes?

2.3 Data Analysis

NVivo8 software was used for qualitative data analysis. The data collected and processed with the above-mentioned software were the recordings related to all the teaching interventions and the simultaneous recording of the teacher's computer screen viewed by the students through the projector. These, once recorded, were transcribed and entered in text form into NVivo8 software, where they were used for the subsequent coding and analysis. More specifically, the produced research material, imported into NVivo8 [17–19] was grouped according to the profile of the students who participated in the research and finally coded according to a series of study categories, which emerged from the literature review and the analysis and recomposition of the produced research material of our teaching intervention. These categories formed the conceptual schema of the analysis of all data generated from this research (Fig. 1).

The first category of Parent Nodes (PN1) refers to the interaction that the teacher had with the students as well as the role he/she played during the teaching interventions. We found that the teacher had a multifaceted role to achieve the best possible result and to be a helper in the assimilation of the material taught by the students.

From the analysis of the available data, we identified five subcategories that formed the Children Nodes (CN) of PN1. More specifically:

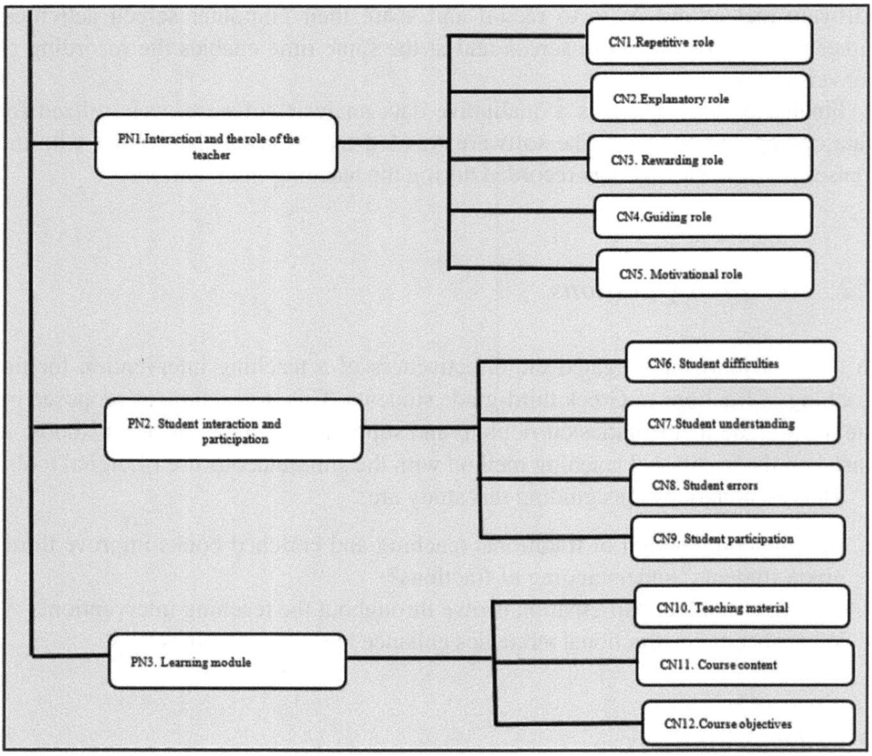

Fig. 1 The categories of analysis

CN1. Repetitive role. Quite often, the teacher appears to repeat the theoretical background of the same or previous lesson, newly explained examples, exercise pronouncements, and solutions to exercises in the context of assisting students' recall of knowledge.

CN2. Explanatory role. In several cases, the teacher had to re-explain concepts already taught and even already solved exercises to confirm that they were clearly understood by the students. This tactic was adopted by the teacher in every case where there was a suspicion or indication that one or some of the students had not correctly grasped the content of the newly acquired knowledge and how to apply it.

CN3. Rewarding role. Throughout the teaching interventions, the teacher did not fail to reward the students for all their correct answers with the aim of both boosting their confidence toward the new knowledge and encouraging them to keep trying.

CN4. Guiding role. In cases where students had difficulty finding a solution to a problem, the teacher, by asking the appropriate questions and making the appropriate suggestions, guided them in finding the tactics to follow to find the right solution.

CN5. Motivational role. Quite often encouraged students who did not take the initiative to participate in the educational process.

The second category of Parent Nodes (PN2) refers to the degree of student interaction and participation in the educational process. After analyzing the data, we distinguished four subcategories that constituted the PN2 Children Nodes, and they are as follows:

CN6. Student difficulties. At several points in the teaching interventions, students had difficulties in understanding and applying the new knowledge, resulting in difficulties in solving the exercises and problems of each teaching unit. In these cases, the teacher's assistance and guidance were intensified to overcome the students' difficulties.

CN7. Student understanding. A record was made of all the correct answers given by the students either verbally to the teacher's questions related to the theory and explanatory activities in the book or in writing about the activities they had to solve.

CN8. Student errors. A record was made of all incorrect answers given by students either verbally during the interaction phases with the teacher or in writing in the activities they were asked to solve.

CN9. Student participation. The students' participation and interaction with each other while doing several activities were very high, which indicates their increased interest and sincere enthusiasm.

The third category of Parent Nodes (PN3) is related to the learning unit taught during the teaching interventions and consists of three subcategories, Children Nodes, which are the following:

CN10. Course material. A record was made of the methods used to teach and promote knowledge about the teaching unit for which the interventions were made.

CN11. Course content. A detailed examination of the learning content of each teaching intervention was conducted.

CN12. Course objectives. We defined the objectives of each chapter taught to students and the expected educational outcomes.

After the qualitative research material was categorized in NVivo software, an external researcher was deemed necessary to perform a coding check to validate the analysis's results internally. In this study, the kappa index appeared with a percentage of more than 90% in most categories, while there were very few cases where it appeared with a percentage of around 70%.

3 Analysis of the Interventions

Making a general summary of students' interaction and participation (PN2) (Table 1) during all interventions, we observed that the teaching intervention in which they experienced the most difficulties (CN6) was the fourth (chapter 25) in which the highest percentage of difficulties occurred with 22.03% while on the contrary, the first teaching intervention (chapter 22) showed the lowest rate of difficulties with 0.23% since this teaching intervention was mainly an introductory lesson with quite easy activities. Regarding concept understanding (CN7), all interventions were around the same percentage (first 12.96%, second 12.28%, fourth 10.23%) except for the third teaching intervention, which showed the lowest percentage of understanding, 6.95%, which was due to specific activities that presented increased difficulty for the students. Remarkable stability was observed in the category of incorrect answers (CN8), with percentages that ranged from 0.69% to 1.81%, which was very satisfactory. Finally, student participation (CN9) was consistently high with similar percentage values, with the highest of all being participation in the third instructional intervention, which reached a value of 68.25%. In the present study, the kappa index appeared with a percentage greater than 90% in most of the categories, while there were a few cases for which the index appeared with a percentage around 70%.

Table 2 gives the percentage change in students' scores between the pre- and post-tests. There is an increase in students' performance in the post-test, which is around 30% for students with high levels of understanding. Of particular importance is the fact that the greatest levels of increase in performance are observed among the weakest students.

3.1 *Role of Students*

Based on participation in the teaching interventions, comprehension of the material taught, and performance in the tests they took, we were able to categorize the students in the classroom into three categories: those who demonstrated high levels

Table 1 Values of PN2's children's nodes

Child nodes	1^{η} Teaching intervention (%)	2^{η} Teaching intervention (%)	3^{η} Teaching intervention (%)	4^{η} Teaching intervention (%)
CN6: Student difficulties	0.23	15.18	11.04	22.03
CN7: Student understanding	12.96	12.28	6.95	10.23
CN8: Student errors	0.69	1.81	1.71	1.79
CN9: Student participation	50.57	46.96	68.25	56.15

Table 2 Percentage change in students' scores on pre- and post-tests

	Pretest (%)	Post-test (%)	Rate of change (%)
Stud 1	74	96	29.7
Stud 2	78	100	28.2
Stud 3	83	100	20.5
Stud 4	61	96	57.4
Stud 5	61	87	42.6
Stud 6	65	100	53.8
Stud 7	26	70	170
Stud 8	35	78	122.8
Stud 9	61	83	36
Stud 10	74	91	22.9
Stud 11	61	91	49.2
Stud 12	0	70	–

of comprehension, moderate levels, and low levels of comprehension (Table 3). In the category of students with high levels of comprehension, we classify the students as Stud. 1, Stud. 2, Stud. 3, Stud. 4, Stud. 10, Stud. 6, and Stud. 11. The main criteria for this distinction were their very good scores in the post-test, which were above 90%, their good scores in the mid-term tests of each module, and finally, their participation during the interventions. We include Stud. 5 and Stud. 9 in the category of students with average levels of comprehension, as their post-test scores were between 80% and 90%, and they performed relatively less than the previous students in the interim tests despite showing high levels of participation.

Stud. 7 and Stud. 8 are classified in the category of students with low levels of understanding as they scored the post-test from 70% to 80%, they had quite low marks on some tests, and their participation was characterized by several errors. In the above categorization, we have not included the Stud.12 student as he is a weak student and has several learning difficulties.

Table 3 Students' performance in the pre-and post-tests and the tests of the teaching interventions

	Pretest (%)	Test 1	Test 2	Test 3	Test 4	Post-test (%)
Stud 1	74	10	7.5	9.5	10	96
Stud 2	78	10	6.5	9.5	5	100
Stud 3	83	9	10	10	10	100
Stud 4	61	10	7.5	10	10	96
Stud 5	61	10	7.5	7	9	87
Stud 6	65	8	10	9.5	6.5	100
Stud 7	26	6	5	4.5	10	70
Stud 8	35	10	1.5	6.5	6.5	78
Stud 9	61	8	10	5.5	7.5	83
Stud 10	74	8	7.5	7	9	91
Stud 11	61	9	9	8	6.5	91
Stud 12	0	2	5	4.5	0	70

3.2 *Role of the Teacher*

During the teaching interventions, the teacher played a very important role as he or she intervened in many ways in the learning process. More specifically, by studying the transcripts, we could see that the role of the teacher was multifaceted, being repetitive, explanatory, rewarding, guiding, and motivating. In each of the four teaching interventions, we observed different percentages of occurrence of each role (Table 4), which is due to the different requirements of each teaching intervention and the different needs of the students in each case.

Table 4 Values of the children's nodes of PN1

Child nodes	1^{η} teaching intervention (%)	2^{η} teaching intervention (%)	3^{η} teaching intervention (%)	4^{η} teaching intervention (%)
CN1: Repetitive role	4.14	2.86	2.42	3.66
CN2: Explanatory role	28.93	21.38	26.97	32.9
CN3: Rewarding role	7.06	8.23	8.32	8.94
CN4: Guiding role	8.17	13.07	11.17	16.19
CN5: Motivational role	14.46	14.43	11.81	12.36

4 Conclusions

The objective of our research design was to validate the perception that teaching fractions with digital tools is the most effective method. This was assessed through pre-tests, post-tests, and interim tests during the interventions. Several researchers, including DeHaan [7], Lee [8], and Orey and Rosa [9], support this view.

We noted a marked improvement in students' knowledge based on their pre- and post-test performances. Analyzing the scores, it was revealed that most students, particularly those with high comprehension levels, showed nearly 30% improvement following the teaching interventions (refer to Table 2). The weakest students demonstrated the most significant improvement in their scores, aligning with Souli [20]. We conducted four tests following each teaching intervention to monitor students' progress. The results indicated a consistent improvement in performance for most students. The results align with the findings of Thambi and Eu [21], Azid et al. [22], and Naidoo and Hajaree [23]. However, the test in Unit 4 posed significant challenges for some students, leading to an anticipated decline in their performance.

The data indicates that teaching interventions integrating traditional methods with digital media have been successful, as evidenced by the students' improved performance and acquisition of new knowledge. The recent teaching intervention significantly enhanced students' performance and effectiveness in handling fractions using digital applications. Through qualitative analysis of the teaching intervention transcripts using NVivo8 software, we examined the profiles and roles of the participating students. Overall, we observed a steady improvement in students' comprehension of new concepts, as evidenced by their increased participation and a concurrent decline in the number of incorrect answers during the interventions. The analysis revealed a decline in the difficulties faced by students, aligning with previous research findings [22–25]. However, slight variations were noted during the fourth instructional intervention, which posed the greatest challenge for students due to its heightened cognitive demands.

Finally, the study examined the teacher's role in the educational process, concluding a qualitative analysis of transcripts using NVivo8 software. Within the scope of this study, which focused on a limited syllabus and timeframe, five roles identified by Rindu and Ariyanti [11] were recognized: repetitive, explanatory, rewarding, guiding, and motivational. The frequency of these roles in each instructional intervention fluctuated based on student needs and the subject matter as they were developed during the intervention. The roles were designed to aid students in acquiring new knowledge and developing metacognitive skills. Clear and explicit guidelines for tasks are crucial, as highlighted by Gillies [13], Gillies and Boyle [14], and Walshaw and Anthony [15]. In this study, the teacher fulfilled these roles through explanatory, instructional, and repetitive methods during all teaching interventions.

Teaching interventions utilizing digital tools positively influence students' comprehension of fractions. Despite this, the study encountered limitations, including a small sample size (12 students) and short intervention duration (8 h). It is recommended that further research be conducted with larger samples and longer interventions to validate these findings.

Overall, it was noted that students' enthusiasm and interest significantly increased with the use of digital tools. The micro-experiments provided a playful and enjoyable method for students to practice fractions. All students engaged actively by seeking answers to each problem displayed on the screen. They interacted by correcting each other, whether accurately or not. Often, students offered satisfactory documentation to support their suggestions. The implementation of interactive enrichment books has notably enhanced the classroom atmosphere. Students found the educational process enjoyable, as they could apply their newfound knowledge to fractions and receive evaluations in an interactive setting. The use of digital tools in the educational process made learning fractions enjoyable and engaging, motivating most students to develop a better understanding and interest.

After concluding this research and considering the results, it would be beneficial to further explore teaching fractions through traditional methods enhanced by digital media. To improve the research, consider replicating it with a larger student sample and incorporating a control group. Additionally, explore enhancing students' teaching and comprehension by utilizing digital games beyond those recommended by the Greek curriculum. Enhancing students' abstraction skills in the early stages of primary school through well-designed digital activities could be beneficial. Investigating the success of this approach is worthwhile.

Acknowledgments This research was partially carried out in the context of the postgraduate thesis of V. Georgoulia.

Disclosure of Interests The authors have no competing interests to declare that are relevant to the content of this article.

References

1. Wu, H.: Basic skills versus conceptual understanding. Am. Educ. **23**(3), 14–19 (1999)
2. Wu, H.: Some remarks on the teaching of fractions in elementary school. https://math.berkeley.edu/~wu/fractions2.pdf (1999). Accessed 10 Oct 2024
3. Nahdi, D.S., Jatisunda, M.G.: Enhancing student's conceptual understanding of fractions through realistic mathematics education. J. Phys. Conf. Ser. **1657**(1), 012047 (2020)
4. Deringöl, Y.: Investigation of primary school students' number sense related to fractions in terms of different variables. Int. Electron. J. Element. Educ. **11**(4), 353–360 (2019). https://doi.org/10.26822/iejee.2019450788
5. Trivena, E., Riyanto, S., Utari, S.: Students' understanding of addition and subtraction of fractions: a case study of Indonesian fifth graders. J. Res. Math. Educ. **6**(3), 267–280 (2017). https://doi.org/10.17583/irem.2017.2389
6. Hariyani, M., Herman, T., Suryadi, D., Prabawanto, S.: Exploration of student learning obstacles in solving fraction problems in elementary school. Int. J. Educ. Methodol. **8**(3), 505–515 (2022)
7. DeHaan, R.L.: The impending revolution in undergraduate science education. J. Sci. Educ. Technol. **14**(2), 253–269 (2005). https://doi.org/10.1007/s10956-005-4425-3

8. Lee, Y.J.: Incorporating digital games into learning: reflections on the educational technology landscape. J. Educ. Technol. Dev. Exchange. **10**(1), 17–28 (2017). https://doi.org/10.18785/jetde.1001.02
9. Orey, M., Rosa, M.: Teaching and learning with technology: effectiveness of ICT integration in schools Indonesian. J. Educ. Res. Technol. **21**(2), 67–83 (2018)
10. Gaggi, O., Ciraulo, F., & Casagrande, M.: Eating pizza to learn fractions. In: Proceedings of the 4th EAI International Conference on Smart Objects and Technologies for Social Good, pp. 220–225 (2018)
11. Rindu, I., Ariyanti, A.: Teacher's role in managing the class during teaching and learning process. Script J. **2**(1), 83–100 (2017)
12. Gillies, R.M., Khan, A.: The effects of teacher discourse on students' discourse, problem-solving, and reasoning during cooperative learning. Int. J. Educ. Res. **47**(6), 323–340 (2008). https://doi.org/10.1016/j.ijer.2008.06.001
13. Gillies, R.M.: Structuring cooperative group work in classrooms. Int. J. Educ. Res. **39**(1–2), 35–49 (2003). https://doi.org/10.1016/S0883-0355(03)00072-7
14. Gillies, R.M., Boyle, M.: Teachers' scaffolding behaviours during cooperative learning. Asia Pac. J. Teach. Educ. **33**(3), 243–259 (2005)
15. Walshaw, M., Anthony, G.: The teacher's role in classroom discourse: a review of recent research into mathematics classrooms. Rev. Educ. Res. **78**(3), 516–551 (2008)
16. Long, C., Dunne, T.: Approaches to teaching primary level mathematics. S. Afr. J. Childhood Educ. **4**(2), 134–153 (2014)
17. Bazeley, P.: Constructing an argument through data analysis. In: Writing Qualitative Research on Practice, pp. 195–205. Brill, London (2009). https://doi.org/10.1163/9789087909086_019
18. Edhlund, B.: NVivo 9 essentials. Lulu.com (2011)
19. Gibbs, G.R.: Qualitative Data Analysis, p. 277. Sage, London (2014)
20. Souli, D.L.: The Contribution of Video Games to the Understanding of Fractions in Primary School Students. Aristotle University of Thessaloniki, Thessaloniki (2021). https://doi.org/10.26262/heal.auth.ir.330074
21. Thambi, N., Eu, L.K.: Effect of students' achievement in fractions using GeoGebra. Sainsab. **16**, 97–106 (2013)
22. Azid, N., Hasan, R., Nazarudin, N.F.M., Md-Ali, R.: Embracing industrial revolution 4.0: the effect of using web 2.0 tools on primary schools students' mathematics achievement (fraction). Int. J. Instr. **13**(3), 711–728 (2020). https://doi.org/10.29333/iji.2020.13348a
23. Naidoo, J., Hajaree, S.: Exploring the perceptions of grade 5 learners about the use of videos and PowerPoint presentations when learning fractions in mathematics. S. Afr. J. Childhood Educ. **11**(1), a846 (2021). https://doi.org/10.4102/sajce.v11i1.846
24. Besalti, M., Kul, U.: Effects of a game-based app on primary students' self-efficacy and achievements in learning fractions during distance education. Turk. Psychol. Couns. Guid. J. **11**(63), 505–520 (2021). https://doi.org/10.17066/tpdrd.1051383
25. Lavidas, K., Georgoulia, V., Komis, V., Voutsina, L.: Instructional intervention for teaching fractions using a digital game environment to third-grade students. Issues Sci. Technol. Educ. **15**, 43–54 (2023)

From Framework to Practice: Competence Development Through Game-Based Experiential Learning in Higher Education

Tharrenos Bratitsis, Marta Ferreira Dias, Isabel Gomes, Gonçalo Meireles, Asta Daunoriene, Yioula Melanthiou, Ewa Rollnik-Sadowska, Ewa Glińska, Audronė Daubarienė, Vasiliki Triga, Marlene Amorim, Ana Luisa Ribeiro, Verónica Kryvohubchenko, and Joana Carrilho

Abstract This chapter discusses how transversal competences can be effectively cultivated via game-based experimental learning in higher education. Grounded in the European Competences Framework for Entrepreneurship framework and learning theories, the ""Game It Away!" programme provides structure in linking competency

---

T. Bratitsis (✉)
Early Childhood Education Department, University of Western Macedonia, Florina, Greece
e-mail: bratitsis@uowm.gr

M. F. Dias · M. Amorim
GOVCOPP, DEGEIT, Universidade de Aveiro, Aveiro, Portugal
e-mail: mfdias@ua.pt

M. Amorim
e-mail: mamorim@ua.pt

I. Gomes · G. Meireles · A. L. Ribeiro
Advancis – Business Services, Lda, Portugal
e-mail: i.gomes@advancis.pt

G. Meireles
e-mail: g.meireles@advancis.pt

A. L. Ribeiro
e-mail: a.ribeiro@advancis.pt

A. Daunoriene · A. Daubarienė
EDU_Lab Center for Excellence in Teaching and Learning, Kaunas University of Technology, Kaunas, Lithuania
e-mail: asta.daunoriene@ktu.lt

A. Daubarienė
e-mail: audrone.daubariene@ktu.lt

Y. Melanthiou · V. Triga
Department of Communication and Marketing, Cyprus University of Technology, Limassol, Cyprus
e-mail: yioula.melanthiou@cut.ac.cy

I. Kazanidis and A. Tsinakos (eds.), *Research on E-Learning and ICT in Education*,
https://doi.org/10.1007/978-3-031-99961-1_3

development with education practice. This programme seeks to engage students in the co-creation of games organized in makerspace environments, emphasizing active learning and real-world application of skills. Empirical pilots revealed positive levels of engagement and competency skills acquired. The chapter further presents alignment issues arising between theoretical models and practical applications; in this respect, it points out that the programme can be differentiated positively as one that proves highly adaptable in as many educational contexts as possible. It also addresses some challenges such as resource availability and facilitator preparedness. Some policy recommendations are extended towards educators, institutions, and policymakers advocating the adoption of experiential and game-driven techniques for enriching the development of transversal competencies.

Keywords Computational thinking · Early childhood education · pedagogical strategies

1 Introduction

In a rapidly changing environment of higher education, the development of transversal competencies has become as important as special discipline knowledge. Nowadays, the labour market requires graduates to have not only domain-specific knowledge but also applied competences that go beyond disciplinary boundaries. Research shows that higher education institutions are under increasing pressure to develop such transdisciplinary competences through innovative pedagogical approaches [1]. The European Competences Framework for Entrepreneurship (EntreComp) represents a significant advance in this field of education, as it offers a comprehensive curriculum for the development of entrepreneurial skills and mindsets necessary for professional success [2]. Despite their widespread recognition, traditional study programme structures often fail to pay sufficient attention to transversal

V. Triga
e-mail: vasiliki.triga@cut.ac.cy

E. Rollnik-Sadowska · E. Glińska
Faculty of Engineering Management, Bialystok University of Technology, Białystok, Poland
e-mail: e.rollnik@pb.edu.pl

E. Glińska
e-mail: e.glinska@pb.edu.pl

V. Kryvohubchenko · J. Carrilho
DEGEIT, Universidade de Aveiro, Aveiro, Portugal
e-mail: veronicak@ua.pt

J. Carrilho
e-mail: joanacarrilho@ua.pt

competences. As a result, a gap between educational outcomes and labour market expectations occurs.

Higher education institutions increasingly admit that traditional teaching methods often fail to engage students in active, experiential learning processes needed to develop transversal competences. Recent studies reveal that university students of different study programmes demonstrate varying levels of proficiency in different transversal competences. For example, complex thinking skills are quite highly developed (3.56 on a standardized scale), while entrepreneurial (2.94) and research (2.99) competences are below the expected theoretical averages [1]. This gap points to the need for didactical changes that can address these challenges through more attractive and effective learning experiences.

The acquisition of transversal competences is complicated due to their interdependent nature. This means that it requires a learning environment that both activates multiple domains of competence and provides meaningful contexts for their application. This complex educational challenge has led researchers to explore alternative didactic frameworks capable of ensuring holistic competence development through immersive, contextually rich learning experiences [3]. Game-based learning has emerged as a particularly promising approach, demonstrating the capacity to support diverse learning outcomes, including meta-skills training, affective development, and systems thinking [3].

The integration of game-based approaches into higher education represents a significant innovation in terms of competence development. For example, the maker space environment creates constructive learning spaces where students can train cognitive functions including all levels of the Bloom taxonomy. Through these immersive experiences, students engage in authentic problem-solving activities within limited spaces. These environments mirror real-world complexity and simultaneously provide structured opportunities for competence development and performance feedback.

Research on game-based learning demonstrates its efficiency across different educational areas, with benefits for developing competencies that traditional instructional methods prove difficult to be effective. For example, board games, specifically, have shown positive results for competence development as they offer hands-on opportunities for students to apply theoretical knowledge in practical scenarios [2]. These gaming environments facilitate experiential learning processes characterized by challenges that are important for transformative learning experiences.

EntreComp provides a comprehensive structure to conceptualize and operationalize the development of entrepreneurial competence [4]. As a well-established reference point for entrepreneurship education, EntreComp offers educators and researchers a coherent framework against which learning interventions can be designed and evaluated [2]. The multidimensional approach of the framework to entrepreneurial competencies aligns well with the complex and multimodal learning experiences that game-based approaches can facilitate.

Systematic literature reviews have examined the application of the EntreComp framework within university settings, highlighting both its conceptual utility and implementation challenge [5]. This growing body of research provides valuable insight into how frameworks such as EntreComp can guide the development of targeted interventions for competency development, particularly when integrated with innovative pedagogical approaches such as game-based experiential learning.

The aim of this chapter is to explore the intersection between theoretical competency frameworks and practical pedagogical approaches through the examination of a game-based experiential learning approach, implemented in higher education settings. Specifically, it investigates how the "*Game It Away!*" project strategy works. The project uses games as important elements for the development of entrepreneurial competence in line with the EntreComp framework [2]. By analysing this innovative educational intervention, we seek to demonstrate effective strategies for transforming abstract competence frameworks into concrete learning experiences that foster measurable skill development.

The significance of this research extends beyond entrepreneurial education to cover broader questions about competence development in higher education. As universities continue to navigate the complex challenge of preparing graduates for the rapidly changing labour market, insights into effective didactic approaches for the development of transversal skills become increasingly valuable. Game-based experiential learning offers a promising route for narrowing the gap between theoretical knowledge and practical competence development, potentially transforming how higher education institutions implement competency-based curriculum.

The chapter is structured as follows; initially the concept of competences in education is discussed, followed by followed by a review of the competence development frameworks. Then the "*Game it Away!*" approach is presented, along with an overview of the outcomes from its implementation. Following, the experiential and game-based learning approaches which were incorporated in this approach are discussed as factors of success for the development of transversal competences, concluding this chapter with theoretical as well as practical contributions and avenues for further research.

2 Understanding Competences in Education

The term competence in educational contexts refers to the combination of knowledge, skills, attitudes, and values that enable individuals to act effectively in complex, real-life situations [6]. The idea of competence as a concept that encompasses more than knowledge dates back to the twentieth century, when educational theories began to focus on practical skills and the overall development of the individual. In the 1970s, the concept of "competence" entered discussions about vocational education, focusing on the specific skills required for success in work environments [7]. At that time, competence was closely linked to performance, approaching it as a set of observable and measurable behaviours. Later, during the 1990s and 2000s, the concept was further developed and expanded, encompassing not only skills but also the knowledge, attitudes, and values required to apply these skills in real-world settings [6]. This broadening of the concept contributed to the adoption of a more holistic view of education, recognizing the importance of personal and social development.

In recent decades, the notion of competence has evolved beyond discipline-specific learning outcomes to encompass transversal, lifelong learning dimensions. This transition mirrors wider shifts in educational policy, labour market demands, and pedagogical approaches that emphasize adaptability, collaboration, and critical thinking [8, 9].

Competence includes four basic dimensions [10]:

1. Knowledge: The information and data that an individual acquires through learning and experience. In the context of education, knowledge is the basis for understanding concepts and processes.
2. Skills: The practical and cognitive abilities that allow the application of knowledge in specific environments.
3. Attitudes: The internal predispositions and beliefs that influence behaviour and decision-making.
4. Values: The moral and social principles that guide human action and contribute to the formation of an individual's identity.

The synthesis of these elements highlights the importance of interdisciplinary and experiential learning in the educational process. In practice, someone is competent in something when they possess basic knowledge in it and can approach it critically, have the required skills to manage this knowledge and produce a result, have the necessary attitudes to judge when and in what contexts they will utilize some of these skills to apply the knowledge and produce a result that is useful for a group of people and finally, are governed by values so as to recognize that the result is produced in an ethical way, correctly, without burdening or harming other people.

In response to globalization, digital transformation, and climate challenges, education systems across the European Union have increasingly adopted competence-based approaches to equip learners with twenty-first-century skills [11]. Competence frameworks serve multiple functions:

- Standardization: Providing shared definitions and expectations across education sectors and national systems [12].
- Transparency: Facilitating the recognition of learning outcomes for mobility and employability [13].
- Accreditation: Enabling alignment with lifelong learning policies and micro-credentialing practices [14].

Frameworks such as key competences for Lifelong Learning and EntreComp operationalize these aims by articulating competences in structured domains. This modular structure allows educators to design curricula and assessment tools aligned with real-world, cross-disciplinary needs.

Traditional curricula have largely prioritized subject mastery, assuming content knowledge naturally leads to transferable abilities. However, empirical studies increasingly demonstrate that transversal competences like collaboration, creativity, and resilience require explicit pedagogical attention [15–17].

Recent frameworks reflect this shift. For instance:

- The LifeComp framework emphasizes personal, social, and learning-to-learn competences such as self-regulation and empathy [18].
- The EntreComp framework defines entrepreneurship as a lifelong competence and identifies 15 sub-skills grouped into "Ideas & Opportunities," "Resources," and "Into Action" [19].
- The DigComp framework focuses on digital literacy, defining competencies related to information literacy, communication, content creation, security, and problem-solving [20].

Notably, these frameworks share a constructivist underpinning: competence is not transmitted but constructed through experience, reflection, and iteration [21, 22]. As a result, educational strategies that promote active, situated, and experiential learning are increasingly recognized as best practice for competence development.

Competency-based learning (CBL) shifts focus from what is taught to what is learned and applied. This implies a reconfiguration of curricula, pedagogy, and assessment, often emphasizing learner agency and formative feedback over summative testing [23].

Key characteristics of CBL include:

- Learning outcomes orientation: Lessons are planned around demonstrable competences, not content delivery.
- Authentic assessment: Students show competence through projects, portfolios, or performances situated in real or simulated contexts [24].
- Personalization: Learning paths accommodate diverse strengths, needs, and aspirations.

The shift also aligns with the growing interest in micro-credentials and digital badges, which promise more flexible and modular recognition of competences acquired inside and outside formal education. These tools are particularly effective in higher education and vocational education contexts where learners seek targeted, and stackable qualifications.

Despite its advantages, implementing competency-based education is not without challenges. Educators often lack concrete tools to assess competences such as critical thinking or initiative. Additionally, school systems constrained by standardized testing may struggle to adopt flexible, student-centred models [25].

Moreover, the assessment of transversal competences presents validity and reliability concerns, as they are context-sensitive and develop over time. Nevertheless, research increasingly points to the value of experiential and game-based approaches as mediators in cultivating such competences [26, 27] a point that will be further explored in subsequent sections of this chapter.

3 Frameworks for Competence Development

The European Union, recognizing the importance of skills development for social and economic cohesion, has developed a series of competence frameworks aimed at shaping policies and strengthening educational practice. In this section, two of the most prominent frameworks, namely EntreComp and DigComp are presented in order to highlight the structure of such theoretical constructs.

3.1 The Entrepreneurship Competence Framework (EntreComp)

The study of entrepreneurship has evolved from a focus on innate personality traits towards a competence-centred approach—observable and measurable sets of knowledge, skills, and attitudes [28]. This paradigm shift recognizes that entrepreneurship can be learned and developed. However, the proliferation of competence models has led to a fragmentation of the field, with few comprehensive and empirically validated frameworks [29]. In this context, the EntreComp, developed by the European Commission [19], has emerged as a significant reference, seeking to provide a common language and a foundation for entrepreneurship education and policy development in Europe and beyond. This review synthesizes recent research that explores, applies, validates, and contextualizes the EntreComp framework, using a selection of published studies.

EntreComp structures entrepreneurial competence into three interconnected areas (Ideas and Opportunities, Resources, Into Action) and 15 specific competences, offering a holistic view of what it means to be entrepreneurial. The need for empirical validation of such frameworks is crucial. Lackéus [30] contributed to this effort by comparing EntreComp with other established models [19, 31], highlighting convergences and divergences and addressing the gap in empirical validation mentioned in previous literature. Understanding the fundamental components of entrepreneurial competence, as explored by Reis et al. [32], provides a broader empirical context for situating and evaluating the structure proposed by EntreComp. Furthermore, systematic reviews and bibliometric analyses like such as [33, 34] have been fundamental in mapping the state-of-the-art research on EntreComp, identifying trends, applications, and gaps in the accumulated knowledge about the framework. Baena-Luna et al. [35] conducted a systematic review highlighting the diversity of contexts in which EntreComp has been applied, while Carlos-Rossi [36] emphasized its utility in deriving structured learning outcomes across various educational levels. These analyses point to a growing body of evidence supporting the framework's versatility, but they also underscore the necessity for continued empirical studies on its effectiveness.

One of EntreComp's areas of greatest impact and research is Entrepreneurship Education (EE). The framework offers a clear structure for curriculum design, teaching, and, crucially, assessment. Recognizing the need for aligned assessment instruments, researchers have developed and validated specific tools. Sánchez Vera and López Vicent [37] created and validated the P5F test, based on EntreComp, to assess entrepreneurial competences in university students. Similarly, Lackéus and Middleton [38] developed detailed assessment rubrics based on EntreComp, providing educators with practical tools to measure student progress. Penaluna and Penaluna [39] also focused specifically on using EntreComp to assess student learning in "enterprise education" programmes.

Beyond assessment, EntreComp is being integrated into pedagogical approaches and technological tools. Arranz et al. [40] analysed the importance of curricular and extracurricular activities in the skills of university students, observing that these activities generate positive results in terms of entrepreneurial skills. Similarly, Daniel et al. [41] found that game-based learning interventions significantly developed entrepreneurial competences aligned with EntreComp dimensions in higher education contexts. Innovative formats like the "Play to Lead" board game [42] are also being explored to foster entrepreneurship-related competencies and leadership in adolescents.

The relevance of EntreComp extends beyond formal education, touching upon broader aspects of entrepreneurial development and labour market readiness. Almeida and Devedzic [43] argue that soft skills, many of which are embedded within EntreComp, are foundational to entrepreneurial success. This aligns with the framework's emphasis on transversal competences such as ethical thinking, motivation, and teamwork. In professional development contexts, Grigg [44] introduced the EntreCompEdu framework, tailored to teacher training, reinforcing the broader pedagogical integration of EntreComp in secondary and vocational education.

Application of the framework has also expanded geographically and across disciplines. Adamec and Marinič [45] explored entrepreneurial competence development among Czech university students, highlighting the framework's adaptability in diverse cultural and educational settings. Pašić et al. [46] similarly examined engineering students in Bosnia and Herzegovina, demonstrating how EntreComp can inform competence development even in technically focused disciplines. Furthermore, Banha et al. [47] critically analysed Portugal's entrepreneurship education policies against EntreComp, revealing alignment gaps and suggesting areas for refinement.

Despite its broad appeal, challenges persist in widespread implementation and in assessing the long-term impact of developing these competences. Bhatti et al. [48] emphasized that without sustained training and reinforcement, the effects of entrepreneurial education on intentions may fade over time. Miço and Cungu [49] further warned that entrepreneurship education remains a complex and challenging process, particularly when not fully embedded in institutional practices. These insights suggest a need for more longitudinal research to evaluate EntreComp's impact on entrepreneurial careers and behaviours over time.

Finally, as entrepreneurship increasingly intersects with other domains such as sustainability and digitalization, EntreComp is being positioned within broader educational paradigms. Moon et al. [50] proposed a dynamic model integrating sustainability into entrepreneurial education using EntreComp as a baseline framework. Similarly, Sánchez Vera and López Vicent [37] explored digital entrepreneurship competence in higher education, emphasizing the relevance of EntreComp in preparing students for evolving digital economies.

Other relevant contributions further enrich the understanding of EntreComp, even if they are less frequently cited in empirical studies. Campbell et al. [51] proposed a bridge between design thinking and EntreComp to structure entrepreneurship workshops, offering an innovative experiential approach to competence development. Gianesini et al. [52] conducted a comparative analysis of entrepreneurial competence models and taxonomies, highlighting convergences with EntreComp and reinforcing its theoretical relevance. In the context of European secondary education, Prendes Espinosa et al. [53] explored teachers' perceptions of the current state and future prospects of entrepreneurial competence, pointing to challenges in its practical implementation. These perspectives broaden the scope of the literature on EntreComp, suggesting important connections between theory, pedagogical innovation, and institutional practice.

In conclusion, EntreComp has established itself as a robust and adaptable framework for understanding and fostering entrepreneurial competence. It bridges theoretical models and practical application across diverse educational and cultural contexts. However, further research is needed to enhance its empirical validation, ensure consistent implementation, and explore its integration with emerging skill domains. Future directions include cross-cultural studies, longitudinal impact assessments, and exploration of its alignment with critical twenty-first-century competencies such as sustainability, leadership, and digital fluency.

3.2 The Digital Competence Framework (DigComp) and Its Extensions

The Digital Competence Framework for Citizens (DigComp) was developed by the European Commission's Joint Research Centre (JRC). It is a key policy instrument designed to foster digital competence across Europe. It was first released in 2013 [17] and it offers a common language and structure for understanding digital skills which are necessary for individuals to actively participate in a digital society. The framework supports policymakers, educators, employers, and individuals in identifying and developing digital competences essential for lifelong learning, employability, active citizenship, and personal development.

The current version, DigComp 2.2 [16], includes updates reflecting technological and societal shifts, such as the growing importance of artificial intelligence, online disinformation, and digital well-being. It outlines five core areas of digital competence:

- Information and data literacy.
- Communication and collaboration.

- Digital content creation.
- Safety.
- Problem-solving.

These areas are detailed into 21 specific competences, each with proficiency levels ranging from Foundation (Level 1) to Highly Specialized (Level 8), aligned with the European Qualifications Framework (EQF). The framework's flexibility and scalability make it suitable for a variety of uses, including national strategies, curricula development, teacher training, and employment services.

Many countries have used DigComp as a foundational structure for their national digital competence frameworks. For instance, Spain's Marco de Competencia Digital and Italy's Repubblica Digitale both align closely with DigComp's descriptors and levels [54, 55]. These national adaptations often serve to localize the framework's recommendations, ensuring cultural and institutional relevance.

One of the most influential adaptations of DigComp is DigCompEdu—the European Framework for the Digital Competence of Educators [56]. While DigComp focuses on general digital literacy for all citizens, DigCompEdu is tailored specifically for teachers and educational professionals. It presents 22 competences across six areas: Professional engagement; Digital resources; Teaching and learning; Assessment; Empowering learners; and Facilitating learners' digital competence.

These competences are structured across six proficiency levels, from Newcomer to Pioneer. The framework aims not only to help educators integrate digital tools into their teaching but also to support them in fostering students' own digital competence.

To facilitate self-reflection and professional development aligned with DigCompEdu, the European Commission developed SELFIEforTEACHERS—a free online self-assessment tool designed to help educators evaluate and improve their digital teaching competence [57]. Launched in 2021 and available in over 30 languages, the tool offers personalized feedback and suggested resources based on the DigCompEdu framework. It has already been adopted by thousands of teachers across Europe, providing valuable data for education systems to monitor digital transformation in schools. Lately, a new version of SELFIEforTEACHERS has been proposed, adapted to Early Childhood Education and Care (ECEC) [58, 59]. It takes into consideration the differences between ECEC and higher educational levels and overall incorporates 31 items of measurement.

DigComp and DigCompEdu have both been widely used in academic research. For example, Caena and Redecker [15] examined how DigCompEdu can be used to design teacher training and professional development programmes, finding that it offers a flexible yet structured approach to digital pedagogy. Similarly, Cabero-Almenara et al. [60] used DigCompEdu to analyse the digital competence of university educators during the COVID-19 pandemic, highlighting significant gaps between perceived and actual competences, especially in digital assessment and student empowerment.

In the context of the labour market, DigComp has been employed to assess the digital skill demands of employers. Kluzer and Pujol Priego [61] analysed job advertisements in Europe using DigComp as a reference, revealing a clear gap between job market expectations and the digital skill levels of many citizens. Their findings support the need for lifelong digital skills training tailored to specific professional contexts.

DigComp has also been central in digital inclusion and citizenship studies. Carretero et al. [20] used the framework to design tools for self-assessment and guidance, particularly for marginalized groups. Such tools are crucial in bridging the digital divide and ensuring equitable access to digital opportunities.

One of the major strengths of DigComp and its derivatives is their ability to adapt to evolving technological contexts. The incremental updates from DigComp 1.0 to 2.2 reflect a responsive approach to emerging trends and educational needs. Additionally, the frameworks' alignment with EQF levels allows for integration into formal education systems and qualification schemes.

In summary, DigComp represents a cornerstone framework for defining and developing digital competence across Europe. Its structured yet adaptable approach makes it useful for a wide range of applications—from educational policy to workplace training and citizen empowerment. The extension into DigCompEdu and tools such as SELFIEforTEACHERS demonstrate the framework's capacity to evolve and address specific needs within the education sector. As digital transformation accelerates, frameworks like DigComp will continue to play a vital role in equipping individuals and societies for the challenges and opportunities of the digital age.

4 Designing the "Game It Away!" Entrepreneurship Programme

Transversal competences such as creativity, collaboration, and resilience are essential to cultivate within formal education. There is an increasing need for innovative, practice-based learning approaches that bridge competence frameworks and real-world application. While frameworks such as EntreComp provide structured guidance for competence development, the challenge remains to create educational experiences that make these competences tangible, actionable, and measurable.

The "*Game It Away!*" Entrepreneurship Training Programme (ETP) was conceptualized precisely to address this gap. By embedding experiential, game-based learning methodologies within the educational process, the programme operationalizes the development of entrepreneurial competences in a dynamic and immersive manner. It aligns closely with the principles outlined in EntreComp, emphasizing not only the acquisition of knowledge but also the application of skills and the shaping of entrepreneurial attitudes.

Building on the structured guidance of EntreComp, the "*Game It Away!*" programme translates theoretical competence frameworks into practical, experiential learning environments, as detailed below.

4.1 The Core Idea: Game Design as a Tool for Teaching Entrepreneurship

Building upon the need for integrated, experiential competence development, the "*Game It Away!*" project leverages game design as a pedagogical strategy to activate entrepreneurial learning processes.

At the foundation of the project is the proposition that game design serves as a dynamic and effective educational tool, enabling learners to engage in complex problem-solving, creative ideation, and iterative development processes—all of which are inherent to entrepreneurial practice [62]. Rather than approaching entrepreneurship education through traditional theoretical instruction, the project adopts a learning-by-doing paradigm, allowing students to experience and navigate entrepreneurial challenges within a structured, low-risk environment [63].

By conceptualizing and developing games—such as board games or escape rooms—students are required to exercise opportunity recognition, critical thinking, teamwork, resource management, and resilience, directly mirroring the key competences outlined in the EntreComp framework [41]. This approach not only deepens cognitive understanding but also cultivates affective and behavioural dimensions of entrepreneurial competence, providing a holistic and transformative learning experience [64].

4.2 Overview of the Structure and Learning Pathway

The "*Game it Away!*" ETP is organized into six interconnected modules, each aligned with a specific stage of entrepreneurial development. The modules follow a logical progression, ensuring a cumulative and coherent learning experience [65]:

- Module I: Let's game it!

  An introductory session establishing fundamental concepts in game design and entrepreneurial thinking, contextualized within the EntreComp framework (3 h).
- Module II: GameCraft: elevate, collaborate, aim for great!

  Focuses on enhancing creative capacity and collaborative ideation through structured brainstorming and user-centred design exercises (3 h).
- Module III: ProtoPlay: master prototypes, budget grooves, creative moves!

  Guides learners through the construction of physical prototypes, emphasizing resource allocation, sustainable practices, and iterative improvement (6 h)
- Module IV: TestCraft: level up your game, learn, and earn!

  Introduces strategies for systematic testing, critical feedback exchange, and adaptive redesign, cultivating resilience and reflective capacity (3 h).
- Module V: PitchCraft: rock the show, steal the glow!

  Develops competencies in strategic communication, persuasive pitching, and audience engagement, essential for entrepreneurial advocacy (4 h).

- Module VI: public presentation and reflection.
  Culminates in public presentation of the final game prototypes to external stakeholders, reinforcing leadership, adaptability, and critical self-assessment (3 h).

The learning pathway reflects the entrepreneurial journey itself, from opportunity identification through product development to market introduction and reflection. Each module builds on the previous one, ensuring a structured yet flexible environment that fosters progressive competence acquisition and practical application of knowledge.

4.3 Makerspaces and Experiential Settings as Enablers of Competence-Building

A distinctive feature of the "*Game It Away!*" programme is the intentional integration of makerspaces as physical and conceptual learning environments. These spaces are instrumental in facilitating hands-on, collaborative, and experimental learning processes, aligning closely with experiential learning theories and constructivist pedagogies.

Within the makerspace environment, students are provided with access to diverse tools, materials, and technologies, enabling them to engage in rapid prototyping, material experimentation, and collaborative problem-solving. The characteristics of makerspaces—openness, community ethos, tolerance for failure, and emphasis on innovation—create a setting that mirrors the dynamic, uncertain conditions of entrepreneurial ecosystems [66].

By situating the learning experience within makerspaces, the programme promotes [67, 68]:

- creativity and innovation: stimulating divergent thinking and novel solution generation.
- collaborative skills: enhancing teamwork, negotiation, and shared decision-making.
- iterative learning: encouraging multiple cycles of testing, failure, feedback, and redesign.
- application of competence frameworks: bridging theoretical models such as EntreComp with lived, embodied learning experiences.

Thus, the makerspace is not merely a physical site but a pedagogical catalyst that transforms entrepreneurial competence development from an abstract goal into a concrete, lived experience. The synergy between game design, experiential settings, and competence frameworks positions the Game it Away! programme as a pioneering model for advancing entrepreneurship education in higher education contexts.

5 Why It Worked: Key Lessons from Pilots and the ETP International Edition

The ETP International Edition provided a platform to enhance entrepreneurial skills through game-based learning and interdisciplinary collaboration. The iterative testing phases of the programme (there were three iterations while developing it) offered valuable insights that contributed to its success in developing important competences for students across various disciplines.

The first notable outcome of the testing phases of the programme was a consistently high level of student engagement. In all three testing rounds, the programme demonstrated consistent outcomes in improving entrepreneurial competences of students. Quantitative surveys found that the average overall satisfaction from students ranges between 4 and 5 out of 5 (see Figs. 1 and 2), which indicates a strong approval of the design and structure of the programme. Both facilitators and students praised the game-based elements for fostering such competences like collaboration and creativity. Additionally, feedback collected from 175 participants across the pilot rounds demonstrated a well-balanced distribution by gender (see Fig. 3) and diverse academic backgrounds, with 47% of students coming from business and management fields (see Fig. 4).

Quantitative data also illustrated significant growth in students' understanding of key entrepreneurial concepts. The ability of the programme to foster learning outcomes was demonstrated by increasing scores from Round 1 to Round 3, particularly in understanding game design, problem-solving, and entrepreneurial thinking. This improvement was seen in students' ratings of their learning outcomes, where the category "Excellent" showed marked growth in the final round. Furthermore, participants were satisfied with their learning experiences, with an average rating of 4.3 out of 5, highlighting the success of the programme in achieving its learning objectives.

Moving on to the analysis of qualitative feedback from students and facilitators, it provided additional depth to the quantitative data. Themes emerged around the

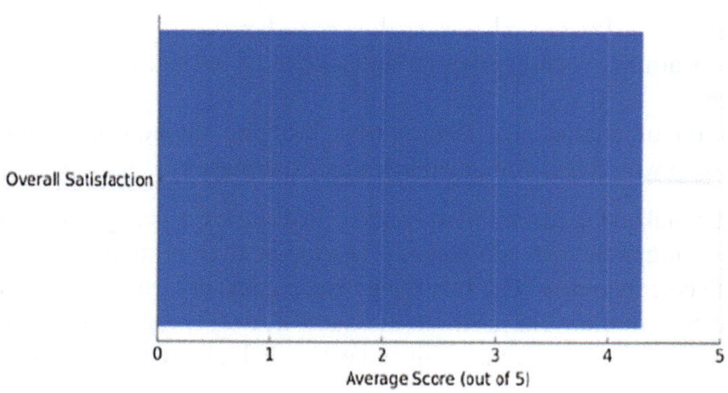

Fig. 1 Overall students' satisfaction of ETP International Edition programme

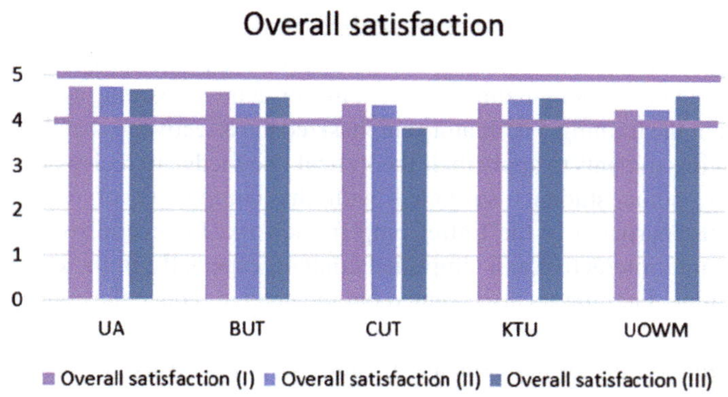

Fig. 2 2 overall students' satisfaction of ETP International Edition programme (by country)

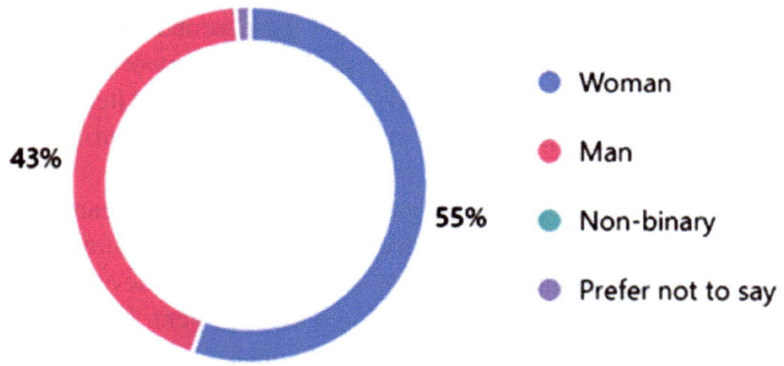

Fig. 3 Gender distribution

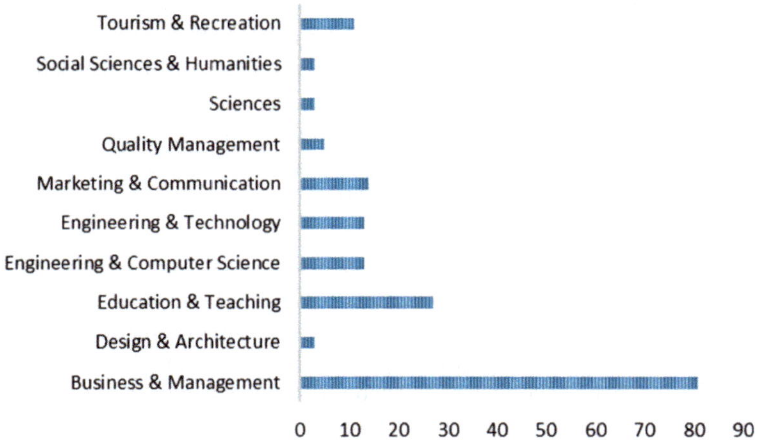

Fig. 4 Academic background

importance of interactive learning, particularly in modules related to game design. Students consistently found the game development and prototyping stages the most engaging, with many recognizing hands-on, maker-centred learning as the most valuable aspect of the training. Facilitators emphasized the effectiveness of these modules in encouraging students to apply their theoretical knowledge to real-world problems.

However, some students struggled with the complexity of certain theoretical concepts, particularly the EntreComp framework. In several rounds, students requested more interactive and simplified content, especially in the early stages of the programme. Facilitators also pointed out the need to reduce the content overload and manage time more efficiently to allow for greater focus on practical application. These insights prompted adjustments to future iterations, such as introducing pre-developed models and case studies to help students better grasp complex theoretical aspects and practices.

A defining feature of the ETP was its incorporation of iterative feedback rounds. Facilitators and students both highlighted how these rounds helped enhance the learning process. Students were encouraged to reflect on their work after each module through learning journals, allowing them to process feedback and iterate on their ideas before final presentations. This process reflected real entrepreneurial action, where continuous improvement and feedback are critical for success.

The game design aspect of the programme was also significant in simulating entrepreneurial action. By engaging students in the creation of game concepts, they were not only learning about the business aspects of game development but also about teamwork, decision-making, and resource management. Facilitators reflected that the gamified elements encouraged critical thinking and problem-solving in a low-risk, collaborative environment. This type of engagement was particularly favourably evaluated as it allowed students to experience the iterative nature of entrepreneurship, where failure is part of the process and leads to better solutions.

One of the key strengths of the ETP International Edition was its interdisciplinary approach. By bringing together students from diverse academic backgrounds—business, engineering, marketing, and education, the programme created a collaborative environment where students could apply different perspectives to solve entrepreneurial challenges. This diversity was highlighted in both qualitative and quantitative feedback, with students reporting that teamwork was one of the most beneficial aspects of the course.

Despite these strengths, some limitations were noted during the pilot testing phases. Time management and content complexity were recurring challenges. For example, students reported that certain modules were too dense and that the timing for hands-on activities, such as prototyping and pitching, was rushed. Feedback suggested that more time should be allocated for practical exercises and that the evaluation processes needed to be simplified. Additionally, the logistical issues around resource accessibility, including difficulties in accessing digital tools and printing, were noted as barriers to efficient learning.

The role of pre-developed materials was instrumental in supporting learning experiences of the students. Facilitators reported that using structured resources—such as pre-prepared models for the pitch sessions—helped streamline certain activities, making them more efficient and less time-consuming. These resources ensured that both students and facilitators had a clear understanding of expectations, reducing confusion, and improving the flow of the programme.

Facilitation played an important role in guiding students through the complexities of entrepreneurial tasks. Facilitators provided mentorship and reflection opportunities, which were vital for helping students refine their ideas and improve their teamwork skills. Facilitator feedback was also crucial in identifying areas where the programme could be improved, especially in terms of simplifying complex tasks and enhancing engagement.

Reflection, both individual and collective, was essential for consolidating the learning process. Structured reflection sessions allowed students to assess their progress and identify areas for improvement. Facilitators emphasized the need for more frequent and focused reflection opportunities to ensure that students could apply their learning in real-world contexts.

The ETP International Edition programme succeeded in fostering entrepreneurial competences by providing students with engaging, hands-on learning experience. The iterative feedback rounds, combined with interdisciplinary collaboration and game-based learning, fostered an environment where students could develop critical thinking, problem-solving, and teamwork skills. However, addressing areas such as time management, content simplification, and logistical support will further enhance the effectiveness of the programme in future iterations.

Through continuous improvement, guided by both student and facilitator feedback, the ETP International Edition has proven to be a dynamic and effective model for entrepreneurial education. By refining content delivery, enhancing interactive elements, and ensuring better resource accessibility, the programme can continue to provide an impactful and immersive learning experience that equips students with the skills necessary to succeed in the entrepreneurial world.

6 The Power of Experiential and Game-Based Learning, for Transversal Competences

One of the key innovations of the "*Game It Away!*" project was its deliberate focus on competence cultivation through immersive, design-led, and game-based experiential learning. Informed by the EntreComp framework and grounded in experiential learning theory, the project positioned learners not only as recipients of knowledge but as co-creators of educational content: specifically, serious games that mirrored real-world entrepreneurial challenges.

At the heart of the "*Game It Away!*" training was an experiential methodology: learners engaged in the full lifecycle of designing educational games on entrepreneurship-related themes. Following Kolb's [21] experiential learning cycle, the programme offered concrete experiences (e.g., building game prototypes), prompted reflective observation (through facilitated debriefing sessions), encouraged abstract conceptualization (via EntreComp-aligned input sessions), and concluded with active experimentation (testing their games with peers or target groups).

In each module, learners were not just learning about entrepreneurship; they were doing entrepreneurship. For instance, in Module 2, participants created rapid game prototypes on decision-making under uncertainty, directly engaging with competences such as coping with ambiguity, uncertainty and risk and taking the initiative, key sub-elements in the EntreComp framework. This approach allowed learners to explore complex content in a dynamic, applied, and creative way. In Module 3, they learned how to listen actively, value others' ideas and mobilize resources in order to create functional prototypes of the games they designed. In Module 4 they practiced collaboration, worked with others and learned through experience by following an adapted Kolb's cycle for testing their prototypes in 3 different stages. In Module 5 they worked creatively in order to pitch their games, among others. Overall, in each one of the modules of the programme, they practiced competences which constitute the EntreComp framework, through hands-on experiences in a realistic context.

Game-Based Learning (GBL) integrates games and corresponding elements into educational contexts to enhance learner engagement and motivation. Rooted in theories of constructivism and experiential learning, GBL provides learners with opportunities to explore complex scenarios, make decisions, and experience consequences in a risk-free environment. Recent studies have highlighted the efficacy of GBL in fostering transversal competences. For example, Pacheco-Velazquez et al. [69] demonstrated that GBL significantly enhances self-directed learning skills among undergraduate students, including self-management, and initiative. Similarly, Fernández-Raga et al. [70] developed a comprehensive process for implementing GBL in higher education, emphasizing its role in promoting critical thinking and adaptability.

Furthermore, GBL aligns with Vygotsky's [71] theory of social constructivism, which underscores the importance of social interaction in cognitive development. Collaborative games encourage learners to communicate, negotiate, and problem-solve collectively, thereby enhancing their interpersonal and teamwork skills.

Unlike passive consumption of games, "*Game It Away!*" invited learners to design games as a medium of learning. This distinction is crucial: learners not only played but constructed the learning process. By developing games targeting real-world entrepreneurial scenarios, such as pitching ideas, managing resources, or handling ethical dilemmas, they were required to internalize and simulate the competences involved.

Game design hence became a pedagogical strategy to elicit transversal skills in action. The learning activities required learners to:

- engage in collaborative design (boosting teamwork and negotiation skills),
- apply systems thinking (structuring rules, objectives, and player interactions),
- iterate based on user feedback (resilience, openness to change), and,
- reflect on value creation and ethical implications (entrepreneurial and civic responsibility).

By aligning each training module with specific EntreComp competences, the project provided a structured yet flexible pathway to transversal competence development. Learners were regularly encouraged to self-assess their growth using simplified EntreComp descriptors. This meta-cognitive dimension (learning to learn) was embedded in reflective checkpoints after each game design phase.

Feedback gathered through evaluation forms and informal debriefs indicated that learners felt more confident in applying entrepreneurial thinking to new contexts. One participant remarked, "*By making a game about pitching, I learned how pitching really works. I wouldn't have understood it as deeply from just reading about it.*"

The "*Game It Away!*" experience underscores that game-based experiential learning is not merely an engaging approach: it is a powerful driver of competence development. Its success was due in part to the coherent integration of design thinking, competence frameworks (EntreComp), and iterative, reflective pedagogy.

Key takeaways for educators and institutions include:

- Pedagogical alignment matters: Linking activities explicitly to competence frameworks helps learners and educators track developmental progress.
- Designing games equals designing learning: Empowering learners to construct games leads to deeper engagement with content and skill application.
- Reflection is essential: Without structured reflection, the link between experience and competence development risks remaining implicit.
- Transversality emerges through complexity: Competences such as creativity, collaboration, and resilience surfaced organically through navigating design complexity.

Looking forward, the "*Game It Away!*" model provides a scalable blueprint for how competence-based learning can be made concrete, playful, and impactful. Whether in entrepreneurship education or broader life skills training, embedding game-based experiential practices offers an inclusive, motivating, and effective route to preparing learners for the challenges of the twenty-first century.

The integration of experiential learning principles within GBL environments creates a synergistic effect, amplifying the development of transversal competences. Games designed with experiential learning cycles encourage learners to engage in concrete experiences, reflect on outcomes, conceptualize strategies, and apply them in subsequent gameplay iterations.

For instance, simulation games that mimic real-world challenges enable learners to experiment with different approaches, receive immediate feedback, and adjust their strategies accordingly. This iterative process not only reinforces content knowledge but also cultivates resilience, adaptability, and critical thinking.

Moreover, the immersive nature of GBL fosters intrinsic motivation, a key factor in sustained engagement and competence development. According to self-determination theory, fulfilling the psychological needs of autonomy, competence, and relatedness enhances motivation. GBL environments inherently support these needs by allowing learners to make choices, experience mastery, and collaborate with peers.

The adoption of experiential and GBL methodologies necessitates a paradigm shift in instructional design and assessment within higher education. Educators must move beyond traditional lecture-based approaches to create interactive, learner-centred environments that prioritize competence development.

Implementing GBL requires careful consideration of learning objectives, game mechanics, and assessment strategies to ensure alignment with desired competences. Additionally, institutions must provide support for faculty development, technological infrastructure, and curricular integration to facilitate the effective adoption of these methodologies.

The fusion of experiential learning and GBL offers a robust framework for cultivating transversal competences in higher education. By engaging learners in meaningful, interactive experiences, educators can better prepare students to navigate the complexities of the modern world.

7 Future Research and Policy Implications

7.1 The Potential for Scaling the Model

The "*Game It Away!*" ETP demonstrates significant potential for scalability across various educational levels (secondary, vocational, and higher education) as well as across different regions. Its modular structure allows for flexibility in implementation, and its makerspace-based approach can be adapted to both technologically advanced and less technologically developed environments. Additionally, future adaptations of the programme into hybrid or fully digital formats (e.g., virtual makerspaces, online game prototyping) could further expand its reach and accessibility.

7.2 Limitations: Cultural Adaptation, Resource Availability, Facilitator Competencies

The proposed ETP is subject to several limitations. One key constraint is the need for cultural adaptation. Concepts of entrepreneurship and approaches to game design can vary significantly across cultures, with cultural norms influencing perceptions of risk, innovation, and failure. Another limitation concerns access to resources, as the availability of makerspaces, prototyping materials, and tools may differ widely between countries and regions. Furthermore, the competencies of facilitators represent a critical limitation. Effective implementation of the programme relies heavily on facilitators' skills in both game design pedagogy and entrepreneurship education.

7.3 *The Role of Micro-Credentials and Competency Frameworks in Model Validation*

The introduction of micro-credentials or digital badges aligned with the dimensions of the EntreComp framework could formalize and validate the competencies acquired through participation in the "*Game It Away!*" ETP. Each module could be linked to specific clusters of competencies, providing recognized evidence of skill acquisition. Micro-credentials could be aggregated into modular certificates, contributing to broader qualifications in innovation, entrepreneurship, or education.

7.4 *Future Work: Longitudinal Studies, Facilitator Training, Cross-Sector Applications*

Future research should include longitudinal studies to evaluate the programme's long-term impact on participants' entrepreneurial attitudes, employability, innovation activities, and resilience (e.g., assessments 1 year and 5 years post-programme completion). It is also necessary to develop and assess formal training programmes for facilitators. Such training should encompass entrepreneurship pedagogy, game-based learning methodologies, makerspace management, and culturally adaptive facilitation strategies.

Beyond formal education, the programme could be adapted for use in corporate innovation training, youth entrepreneurship initiatives, social entrepreneurship laboratories, and capacity-building for non-governmental organizations. Future studies should explore and evaluate the programme's effectiveness across these sectors.

Moreover, further research should examine the effectiveness of a hybrid version of the "*Game It Away!*" programme, incorporating virtual makerspaces, online prototyping tools, and remote collaboration platforms.

8 Conclusions

This chapter has examined how transversal competences—such as creativity, collaboration, resilience, and initiative—can be effectively cultivated through a pedagogically structured, game-based experiential learning approach. Anchored in the EntreComp framework and accordant with the constructivist and experiential learning perspectives [41, 66], the "*Game It Away!*" programme maps the abstract competency models mentioned in the original model to tangible learning interventions. Eventually, these interventions will become both measurable as well as scalable.

The work presented herein advances educational theory in at least three areas. First, it has taken one step in actualizing CBL by embedding the framework into a practical, iterative, and student-led approach. Different from the majority of

CBL models, this approach grounds the development of the competences from the learners' lived experiences of competences being developed in a first-hand, hands-on fashion. Second, the chapter broadens the scope of GBL as a study that is more than content transmission and engagement. It positions game design as purposeful meta-cognitive activity: besides imparting knowledge, it builds entrepreneurship identity and behaviour. This concurs with the ongoing literature establishing that GBL is a major intermediary in meta-skills and transversal learning [3, 64, 65]. Thirdly, the chapter builds on discussions about the learning of EntreComp within a nonformal, experientially sculpted physical layout such as a makerspace through a hybrid, informal, and adaptive environment that fosters skill authentic development [61–63].

The "Game It Away!" model practical contributes by providing a tried and tested template for institutions willing to overhaul curricula on the basis of transversal competences. The modular nature of the ETP, the alignment of that training with the competence frameworks, and the maker-centred iterative learning process design make it applicable to numerous different institutional settings and learner profiles. It attempts to provide solutions to the known practical challenges when implementing CBL, supporting formative assessment (through reflective checkpoints), increase motivation (game co-creation), and lend relevance to reality (by simulating that very act of being an entrepreneur). Moreover, for teaching professionals, it can provide a framework not just to teach but to encourage co-construction of learning experiences. Institutions can use it as a means to mainstream transversal competence development while retaining their curricular strengths. Policymakers could consider inserting EU-level frameworks, including EntreComp and DigComp, into national or institutional policies through pedagogically sound procedures.

The "*Game It Away!*" model has effectively presented how transversal competences are best learned through experiential, game-based pedagogies. Through this, great harmonization can take place between actual framework of what is taught and pragmatic implementation of the knowledge from passive instruction to one that is achieves active, reflective, and competence-rich engagement. As the global higher education sector is facing an urgent need to prepare graduates for the highly complex and unpredictable future environment, educators, institutions, and policymakers ought to embrace this model as a genuine and necessary approach to relevant, interesting, and significantly impactful educational experiences.

Acknowledgments This work is funded by the Erasmus+ programme, key action 2, Cooperation partnerships in higher education (Proj No: 2022-1-PT01-KA220-HED-000087026).

References

1. Guerra-Macas, Y., Tobón, S.: Development of transversal skills in higher education programs in conjunction with online learning: relationship between learning strategies, project-based pedagogical practices, e-learning platforms, and academic performance. Heliyon. **11**(2) (2025)

2. Dias, M.F., Alves, C., Amorim, M.: Board games and entrepreneurial skills acquisition: A mixed-methods study on the efficacy of the EntreComp framework in higher education. In: The 18th European Conference on Games Based Learning, pp. 275–280. Academic Conferences and Publishing Limited, Manchester (2024). https://doi.org/10.34190/ecgbl.18.1.2666
3. Baalsrud Hauge, J.: Current competencies of game facilitators and their potential optimization in higher education: multimethod study. JMIR Ser. Games. **9**(2), e25481 (2021)
4. McCallum, E., Weicht, R., McMullan, L., & Price, A.: EntreComp into Action-Get Inspired, Make it Happen: A User Guide to the European Entrepreneurship Competence Framework (No. JRC109128). Joint Research Centre (2018)
5. Osipovskaya, E., Coelho, A., & Tasi, P.: Literature review on the application of EntreComp framework in university settings. EDULEARN24 Proceedings, pp. 3626–3632 (2024)
6. Rychen, D.S., Salganik, L.H. (eds.): Key Competencies for a Successful Life and a Well-Functioning Society. Hogrefe & Huber Publishers, Newburyport (2003)
7. Boyatzis, R.E.: The Competent Manager: A Model for Effective Performance. John Wiley & Sons, New York (1982)
8. Deakin Crick, R.: Key competencies for education in a European context: Narratives of accountability or care. Eur. Educ. Res. J. **7**(3), 311–318 (2008). https://doi.org/10.2304/eerj.2008.7.3.3
9. Deakin Crick, R., Stringher, C., Ren, K.: Learning to Learn. International Perspectives from Theory and Practice. Routledge, London (2014). https://doi.org/10.4324/9780203078044
10. Weinert, F.E.: Concept of competence: A conceptual clarification. In: Rychen, D.S., Salganik, L.H. (eds.) Defining and Selecting Key Competencies, pp. 45–66. Hogrefe & Huber, Seattle (2001)
11. European Commission: Proposal for a Council Recommendation on Key Competences for Lifelong Learning. Brussels (2018)
12. Mulder, M. (ed.): Competence-Based Vocational and Professional Education. Bridging the Worlds of Work and Education. Springer, Cham (2017)
13. Cedefop: transparency and transferability of learning outcomes: a 20-year journey: analysis of developments at European and national level. Cedefop research paper. Publications Office of the European Union. https://doi.org/10.2801/3413866 (2024)
14. OECD: Micro-Credential Innovations in Higher Education: Who, What and why? OECD Education Policy Perspectives, vol. 39. OECD Publishing, Paris (2021). https://doi.org/10.1787/f14ef041-en
15. Caena, F., Redecker, C.: Aligning teacher competence frameworks to 21st century challenges: the case for the European digital competence framework for educators (Digcompedu). Eur. J. Educ. **54**, 12345 (2019). https://doi.org/10.1111/ejed.12345
16. Vuorikari, R., Kluzer, S., Punie, Y.: DigComp 2.2: the Digital Competence Framework for Citizens - with New Examples of Knowledge, Skills and Attitudes. Publications Office of the European Union, Luxembourg (2022). https://doi.org/10.2760/115376
17. Ferrari, A.: DIGCOMP: A Framework for Developing and Understanding Digital Competence in Europe. Publications Office of the European Union, Luxembourg (2013). https://doi.org/10.2788/52966
18. Sala, A., Punie, Y., Garkov, V., Cabrera Giraldez, M.: LifeComp: the European Framework for Personal, Social and Learning to Learn Key Competence. Publications Office of the European Union, Luxembourg (2020). https://doi.org/10.2760/302967
19. Bacigalupo, M., Kampylis, P., Punie, Y., Van den Brande, G.: EntreComp: The Entrepreneurship Competence Framework. Publication Office of the European Union, Luxembourg (2016). https://doi.org/10.2791/593884
20. Carretero, S., Vuorikari, R., Punie, Y.: DigComp 2.1: The Digital Competence Framework for Citizens with Eight Proficiency Levels and Examples of Use. Publications Office of the European Union, Luxembourg (2017)
21. Kolb, D.: Experiential Learning: Experience as the Source of Learning and Development. Prentice Hall, Englewood Cliffs (1984)
22. Illeris, K.: An overview of the history of learning theory. Eur. J. Educ. **3**(53), 86–101 (2018)

23. Torenbeek, M., Jansen, E., Hofman, A.: The effect of the fit between secondary and university education on first-year student achievement. Stud. High. Educ. **35**(6), 659–675 (2010)
24. Kunnari, I., Ilomäki, L.: Reframing teachers' work for educational innovation. Innov. Educ. Teach. Int. **53**(2), 167–178 (2016)
25. van der Klink, M., Boon, J., Schlusmans, K.: Competences and vocational higher education: now and in future. Eur. J. Voc. Train. **40**, 219 (2007)
26. Iatraki, G., Mikropoulos, T.A.: Augmented reality in physics education: students with intellectual disabilities inquire the structure of matter. PRESENCE Virt. Aug. Real. **31**, 89–106 (2022). https://doi.org/10.1162/pres_a_00374
27. Bellotti, F., Kapralos, B., Lee, K., Moreno Ger, P., Berta, R.: Assessment in and of serious games: an overview. Adv. Human-Comput. Interact. (2013). https://doi.org/10.1155/2013/136864
28. Chouhan, V.S., Srivastava, S.: Understanding competencies and competency modeling — a literature survey. IOSR J. Bus. Manag. **16**(1), 14–22 (2014). https://doi.org/10.9790/487X-16111422
29. Morris, M., Webb, J., Singhal, S., Fu, J.: A competency-based perspective on entrepreneurship education: conceptual and empirical insights. J. Small Bus. Manag. **51**(3), 352–369 (2013). https://doi.org/10.1111/jsbm.12023
30. Lackéus, M.: Comparing the impact of three different experiential approaches to entrepreneurship in education. Int. J. Entrepreneur. Behav. Res. **26**(5), 937–971 (2020). https://doi.org/10.1108/IJEBR-04-2018-0236
31. Bartram, D.: The great eight competencies: a criterion-centric approach to validation. J. Appl. Psychol. **90**(6), 1185–1203 (2005). https://doi.org/10.1037/0021-9010.90.6.1185
32. Reis, D., Leme Fleury, A., Monteiro de Carvalho, M.: Consolidating Core entrepreneurial competences: towards A meta-competence framework. Int. J. Entrepreneur. Res. **27**(1), 179–204 (2020). https://doi.org/10.1108/IJEBR-02-2020-0079
33. Pérez-López, M.C., González-López, M.J., Rodríguez-Ariza, L.: EntreComp framework: a bibliometric review and research trends. Sustainability. **15**(2), 1285 (2023). https://doi.org/10.3390/su15021285
34. Secundo, G., Rippa, P., Cerchione, R.: Digital academic entrepreneurship: a structured literature review and avenue for a research agenda. Technol. Forecast. Soc. Change. **157**, 120118 (2020). https://doi.org/10.1016/j.techfore.2020.120118
35. Baena-Luna, P., García-Río, E., Monge-Agüero, M.: Entrecomp: marco competencial para el emprendimiento. Una revisión sistemática de la literatura sobre su uso y aplicación. CIT Informacion. Tecnologica. **31**(2), 163–172 (2020). https://doi.org/10.4067/s0718-07642020000200163
36. Carlos-Rossi, A.: Entrepreneurial education: Competencies and learning outcomes. A systematic review. In: 2021 IEEE 1st International Conference on Advanced Learning Technologies on Education & Research (ICALTER). IEEE, New York (2021). https://doi.org/10.1109/ICALTER54105.2021.9675118
37. Sánchez Vera, M.M., López Vicent, P.: The competence of digital entrepreneurship in education: analysis of the perception of university students. Intangible Capital. **20**(2), 376 (2024). https://doi.org/10.3926/ic.2417
38. Lackéus, M., Middleton, K.W.: Assessing experiential entrepreneurship education: key insights from five methods in use at a venture creation programme. In: Hyams-Ssekasi, D., Caldwell, E. (eds.) Experiential Learning for Entrepreneurship, pp. 19–49. Palgrave Macmillan, Cham (2018). https://doi.org/10.1007/978-3-319-90005-6_2
39. Penaluna & Penaluna: In search of entrepreneurial competencies: peripheral vision and multidisciplinary inspiration. Ind. High. Educ. **35**(8), 095042222096379 (2018). https://doi.org/10.1177/0950422220963796
40. Arranz, N., Ubierna, F., Arroyabe, M.F., Perez, C., Fdez Arroyabe, J.C.: The effect of curricular and extracurricular activities on university students' entrepreneurial intention and competences. Stud. High. Educ. **42**(11), 1979–2008 (2017). https://doi.org/10.1080/03075079.2015.1130030

41. Daniel, A.D., Negre, Y., Casaca, J., Patrício, R., Tsvetcoff, R.: The effect of game-based learning on the development of entrepreneurial competence among higher education students. Educ. Train. (2024). https://doi.org/10.1108/et-10-2023-0448
42. Bonaiuto, F., Perucchini, P., Placidi, V., Faggioli, S., Barroca, A., Ferot, C., Netz, L., Xie, M., Bonaiuto, M.: “Play to Lead” board game as A potential intervention to promote entrepreneurship competences and servant leadership skills in European adolescents. Psychol. Hub. **40**(2), 59–68 (2023). https://doi.org/10.13133/2724-2943/17761
43. Almeida, F., Devedzic, V.: The relevance of soft skills for entrepreneurs. J. East Eur. Manage. Stud. **27**(1), 157–172 (2022). https://doi.org/10.5771/0949-6181-2022-1-157
44. Grigg, R.: EntreCompEdu, a professional development framework for entrepreneurial education. Educ. Train. **63**(7/8), 1058–1072 (2021). https://doi.org/10.1108/et-06-2020-0166
45. Adamec, P., Marinič, P.: Entrepreneurial competences development of university students in the Czech Republic. Entrepreneur. Innov. **19**(1), 12–19 (2024). https://doi.org/10.34190/ecie.19.1.2561
46. Pašić, M., Vatreš, A., Bijelonja, I., Pašić, M.: Analysis of development of entrepreneurship competences of engineering students based on EntreComp framework. Int. J. Ind. Eng. Manage. **14**(2), 162–181 (2023). https://doi.org/10.24867/IJIEM-2023-2-331
47. Banha, F., Graça, A.R., Góis, B., Banha, F.M.: A critical analysis of the Portuguese framework for entrepreneurship education considering the EntreComp competence framework. Entrep. Educ. **8**(1), 1–28 (2025). https://doi.org/10.1007/s41959-025-00135-y
48. Bhatti, M.A., Al Doghan, A., Mat Saat, S.A., Juhari, A.S., Alshagawi, M.: Entrepreneurial intentions among women: does entrepreneurial training and education matters? (pre- and post-evaluation of psychological attributes and its effects on entrepreneurial intention). J. Small Bus. Enterp. Dev. **28**(2), 167–184 (2021). https://doi.org/10.1108/jsbed-09-2019-0305
49. Miço, H., Cungu, J.: Entrepreneurship education, a challenging learning process towards entrepreneurial competence in education. Admin. Sci. **13**(1), 22 (2023). https://doi.org/10.3390/admsci13010022
50. Moon, C.J., Walmsley, A., Apostolopoulos, N.: Enterprise and entrepreneurship Education: Exploring a new dynamic model to integrate sustainability within curricula. Entrepreneur. Innov. **18**(2), 1056–1060 (2023). https://doi.org/10.34190/ecie.18.2.1650
51. Campbell, D., Balis, G., Aryana, B.: Bridging design thinking and entrecomp for entrepreneurship workshops: A learning experience. In: Proceedings of the Sixth International Conference on Design Creativity (ICDC 2020), pp. 247–254. The Design Society, Glasgow (2020). https://doi.org/10.35199/ICDC.2020.31
52. Gianesini, G., Cubico, S., Favretto, G., Leitão, J.: Entrepreneurial competences: comparing and contrasting models and taxonomies. In: Entrepreneurship and the Industry Life Cycle, pp. 13–32. Springer International Publishing, Cham (2018). https://doi.org/10.1007/978-3-319-89336-5_2
53. Prendes Espinosa, M.A.P., Solano Fernandez, I.M.A., Gonzalez Martinez, J., Cerdan Cartagena, F.: Competencia de emprendimiento en educación secundaria: percepción del profesorado sobre el estado actual y las posibilidades futuras en el contexto europeo. Bordón Rev. Pedag. **72**(2), 153–172 (2020). https://doi.org/10.13042/bordon.2020.67626
54. INTEF: Marco de referencia de la competencia digital docente. Instituto Nacional de Tecnologías Educativas y de Formación del Profesorado (2020)
55. AgID: Repubblica Digitale: Strategia nazionale per le competenze digitali. Agenzia per l’Italia Digitale (2021). https://repubblicadigitale.innovazione.gov.it/
56. Redecker, C.: European framework for the digital competence of educators: DigCompEdu. Publications Office of the European Union (2017). https://doi.org/10.2760/159770
57. European Commission.: SELFIEforTEACHERS: Self-Reflection on Effective Learning by Fostering the Use of Innovative Educational Technologies (2022). https://education.ec.europa.eu/selfie-for-teachers
58. Kontovourki, S., Bratitsis, T., Economou, A., Kapsalis, G.: SELFIEforTEACHERS for early childhood education and care professionals: adaptation and piloting. In: ICERI2024 Proceedings, pp. 5824–5831. IATED, Valencia (2024)

59. Economou, A., Kapsalis. G., Kontovourki, S., Bratitsis, T.: Adapting SELFIEforTEACHERS self-reflection tool to support professional development of (ECEC) professionals. In: 31st EECERA Annual Conference - Children's Curiosity, Agency and Participation: Challenges for Professional Action and Development, pp. 55–56 (2023)
60. Cabero-Almenara, J., Romero-Tena, R., Palacios-Rodríguez, A.: The digital competence of university teachers: A study based on the DigCompEdu framework. Educ. Inf. Technol. **26**(4), 4819–4832 (2021). https://doi.org/10.1007/s10639-021-10565-7
61. Kluzer, S., & Pujol Priego, L.: DigComp into action: get inspired, make it happen. JRC Science for Policy Report (2018). https://doi.org/10.2760/112945
62. Gatti Jr, W., Kim, B.: Gameplay and game design to enhance design thinking in entrepreneurship education. In: Kalir, R., Filipiak, D. (eds.) Proceedings of the 2019 Connected Learning Summit, pp. 54–61 (2019)
63. Fergusson, L.: Learning by… Knowledge and skills acquisition through work-based learning and research. J. Work Appl. Manage. **14**(2), 184–199 (2022). https://doi.org/10.1108/JWAM-12-2021-0065
64. Yang, Q., Zhang, Y., Lin, Y.: Study on the influence mechanism of virtual simulation game learning experience on student engagement and entrepreneurial skill development. Front. Psychol. **12**, 772157 (2022). https://doi.org/10.3389/fpsyg.2021.772157
65. Bratitsis, T., Ferreira Dias, M., Gomes, I., Meireles, G., Ryciuk, U., Daunoriene, A., Melanthiou, Y., Rollnik-Sadowska, E., Daubarienė, A., Triga, V., Amorim, M., Ribeiro, A.L., Silva, C., Alves, C.: Interweaving EntreComp, game-based learning and experiential learning for the design of an entrepreneurship training programme. In: Proceedings INTED2025, pp. 6909–6918. IATED, Valencia (2025)
66. Hui, J., Gerber, E.: Developing makerspaces as sites of entrepreneurship. In: Proceedings of the 2017 ACM Conference on Computer Supported Cooperative Work and Social Computing (CSCW'17), pp. 2023–2038. ACM, New York (2017). https://doi.org/10.1145/2998181.2998264
67. Li, J., Eberhart, R., Eesley, C.: Makerspaces, entrepreneurship, and institutions. SSRN Working Paper (2022). https://doi.org/10.2139/ssrn.4046207
68. Soomro, S., Casakin, H., Nanjappan, V., Georgiev, G.: Makerspaces fostering creativity: A systematic literature review. J. Sci. Educ. Technol. **32**(3), 530–548 (2023). https://doi.org/10.1007/s10956-023-10041-4
69. Pacheco-Velazquez, E., Rodés, V., Salinas-Navarro, D.: Developing learning skills through game-based learning in complex scenarios: A case in undergraduate logistics education. J. Technol. Sci. Educ. **14**(1), 169–183 (2024). https://doi.org/10.3926/jotse.2219
70. Fernández-Raga, M., Aleksić, D., İkiz, A.K., Markiewicz, M., Streit, H.: Development of a comprehensive process for introducing game-based learning in higher education for lecturers. Sustainability. **15**(4), 3706 (2023). https://doi.org/10.3390/su15043706
71. Vygotsky, L.S.: Mind in Society: the Development of Higher Psychological Processes. Harvard University Press, Cambridge (1978)

Fostering Early Childhood Science Education Through Visual Programming: A Case Study with Pre-service Teachers Using ScratchJr

Georgios K. Zacharis

Abstract This study underscores the considerable potential of integrating ScratchJr into Early Childhood Science Education (ECSE), particularly within the context of Greek pre-service early childhood teacher (ECT) training. The findings illustrate how pre-service ECTs engage with ScratchJr to design science-oriented learning activities, highlighting the pedagogical value of structured, tutorial-based approaches alongside more creative formats such as storytelling and game-based learning. These instructional trends reflect a combination of goal-directed teaching methods and interactive, narrative-driven strategies. Although participants exhibited varied levels of programming proficiency, a substantial number demonstrated strong competence in using ScratchJr, indicating increased confidence in employing visual programming. Through the development of curriculum-aligned projects, pre-service ECTs enhanced both their scientific content knowledge and digital literacy. In alignment with the Technology Acceptance Model (TAM), the study revealed high levels of perceived usefulness and ease of use among pre-service ECTs, further supporting their intention to adopt ScratchJr in their future teaching practices. Ultimately, the study advocates for the integration of visual programming environments in ECTs education programs to cultivate a generation of technologically adept educators capable of fostering scientific inquiry and innovation in early childhood classrooms.

Keywords Early Childhood Science Education · Computational Thinking · Visual Programming · ScratchJr · Self-efficacy · TAM

G. K. Zacharis (✉)
Aristotle University of Thessaloniki, Thessaloniki, Greece
e-mail: gzacharis@nured.auth.gr

I. Kazanidis and A. Tsinakos (eds.), *Research on E-Learning and ICT in Education*,
https://doi.org/10.1007/978-3-031-99961-1_4

1 Introduction

Science in Early Childhood Education (ECE) plays a foundational role in fostering children's curiosity, observational skills, and basic understanding of the natural world. From an early age, children exhibit a natural inclination to explore their environment, ask questions, and engage with their surroundings through hands-on experiences—activities inherently aligned with the nature of scientific inquiry [1]. Consequently, contemporary early childhood pedagogy emphasizes the integration of science into preschool curricula as a means of nurturing inquiry-based learning and exploration [2].

A substantial body of research underscores the importance of introducing scientific experiences during the early years, particularly for children aged 3–5. Early engagement with science not only promotes positive attitudes toward the subject but also lays the conceptual groundwork essential for future academic success. It is therefore pedagogically imperative to acknowledge and build upon the prior knowledge and experiential background that children bring with them into educational settings. Encouraging and supporting their innate curiosity contributes significantly to the development of scientific literacy from the earliest stages of learning [3].

Despite the well-documented benefits of ECSE, a key challenge lies in the preparation of ECTs. Specifically, the development of pre-service ECTs' pedagogical content knowledge and self-efficacy in delivering science instruction remains limited. ECTs who have not had adequate exposure to science content or related teaching methodologies during their undergraduate education often express discomfort and lack confidence in teaching science [4]. This gap in preparation can negatively impact the quality of instruction and may result in early disengagement from science among young learners—a trend that often persists into later stages of education [1].

In response to these challenges, Educational Robotics (ER) has emerged as a promising pedagogical approach within STEM education (Science, Technology, Engineering, and Mathematics). ER involves engaging students in programming robots to perform specific tasks, thereby promoting hands-on, inquiry-based learning [5, 6]. Rooted in constructivist and constructionist theories, ER facilitates the development of twenty-first-century skills, including active participation, collaboration, problem-solving, and computational thinking, by allowing students to observe and manipulate robot behavior in response to their programming inputs [7, 8]. Research has increasingly demonstrated the potential of ER to enhance motivation, facilitate STEM-related learning, and support the development of transversal skills [6].

Central to Educational Robotics (ER) is the development of Computational Thinking (CT)—a set of cognitive processes first introduced by Wing [9], which includes skills such as algorithmic reasoning, problem decomposition, and data representation. CT has since been recognized as a foundational literacy skill, such as reading, writing, and arithmetic. In the context of early childhood education, the introduction of CT is considered developmentally appropriate and pedagogically valuable, as it fosters problem-solving and reasoning abilities while also laying the groundwork for future engagement with computer programming [10]. Empirical research supports the notion that tools designed specifically for young learners—such as robots featuring visual or tangible programming interfaces—can effectively cultivate CT skills in early learning environments [11–14]. These developmentally appropriate technologies promote active, hands-on engagement and align closely with the natural curiosity and exploratory behavior observed in early learners.

Among the tools designed to facilitate CT development, ScratchJr stands out as a prominent visual programming environment tailored for children aged 5–7. With its block-based interface and intuitive design, ScratchJr allows young learners to create interactive stories and games without requiring reading or prior coding experience. This environment enables young children to sequence commands, debug errors, and visualize abstract computational processes, thereby enhancing their cognitive and creative capacities [15, 16]. For educators, ScratchJr provides an opportunity to integrate digital literacy into the curriculum in a developmentally appropriate and pedagogically meaningful way.

Research on ScratchJr highlights its efficacy not only in promoting core CT concepts—such as sequencing, symbolic reasoning, and goal-directed programming—but also in fostering creativity, problem-solving, and higher-order thinking skills [17–19]. Moreover, ScratchJr has been shown to support active learning, positioning children as content creators rather than passive consumers of digital media [20, 21]. A systematic review by Louka [22] further identifies ScratchJr as one of the most effective tools for introducing coding and CT in early educational settings, with numerous studies reporting positive learning outcomes.

In light of these findings, it is essential that pre-service ECTs not only develop the technical and conceptual foundations of CT, but also foster a positive disposition and confidence in integrating programming tools into their instructional practices. Research conducted in Greece by Kalogiannakis and Papadakis [23] demonstrated a significant increase in the self-efficacy of 122 pre-service ECTs in CT following their engagement with ScratchJr. Additionally, participants expressed positive perceptions regarding both the ease of use and the perceived usefulness of ScratchJr, indicating a strong willingness to incorporate it into their future teaching practices. Furthermore, findings by Shina et al. [8] indicate that pre-service teachers' acceptance and self-efficacy regarding ER significantly improved following the completion of an ER training program. Notable differences were observed between the beginning and end of the course, particularly in terms of perceived ease of use, enjoyment, attitudes, and overall self-efficacy related to ER integration.

Providing training to pre-service ECTs in environments such as ScratchJr represents a critical step toward the meaningful integration of technology and science education in a developmentally appropriate manner. Through the design and implementation of science-oriented projects using ScratchJr, future ECTs acquire practical experience in leveraging technology to promote scientific literacy, foster creativity, and increase learner engagement in early childhood settings.

2 Methods

2.1 *Purpose*

The purpose of the current study is to examine the impact of using ScratchJr on pre-service ECTs' self-efficacy in Computational Thinking (CT), as well as their acceptance of ScratchJr as a tool for learning and teaching both CT and science concepts. The study aims to explore the role of visual programming environments in preparing future ECTs to integrate digital tools into developmentally appropriate science education. The research was guided by the following questions:

RQ1: How do pre-service early childhood teachers utilize ScratchJr to design science-related educational projects, and what patterns emerge in terms of content selection, project format, and programming complexity?

RQ2: To what extent does the use of ScratchJr enhance pre-service early childhood teachers' self-efficacy in applying in future science education practices?

RQ3: To what extent do pre-service early childhood teachers accept ScratchJr as a tool for learning and teaching Computational Thinking within the context of science education?

2.2 *Context and Participants*

The study was conducted within the framework of the undergraduate course "Interdisciplinary Approaches to Science and Information and Communication Technologies (ICT) in Education", offered by the Department of Early Childhood Education at the Aristotle University of Thessaloniki, Greece. The course spans one academic semester and focuses on equipping future ECTs with the skills to integrate science learning with age-appropriate digital tools, with particular emphasis on ScratchJr. A total of 45 pre-service ECTs participated in the study, of whom 42 (93.3%) identified as female. Most participants had no prior experience with visual programming environments, and none had used ScratchJr before enrolling in the course.

2.3 *Instructional Design*

The instructional approach combined theoretical input with practical application. During the first half of the semester, pre-service ECTs were introduced to ScratchJr as an educational tool through a series of guided tutorials, collaborative tasks, and hands-on activities. Core programming concepts—including sequencing, loops, and event-based actions—were introduced using storytelling and visual communication techniques suitable for young learners. Pre-service ECTs were then assigned to design interactive projects that integrated ECSE topics such as states of matter, living organisms, and Earth systems. These themes were aligned with the national early childhood curriculum in Greece. Each pre-service ECT created a digital tutorial, interactive story, or simple game in ScratchJr to convey a scientific concept in a playful and pedagogically meaningful manner.

2.4 *Project Development and Support*

Project development occurred over several weeks and was supported through structured in-class planning sessions, iterative peer feedback, and individual guidance from the course instructor. In the final 2 weeks of the semester, pre-service ECTs presented their projects in plenary sessions, which included discussion and constructive critique from peers and faculty members.

2.5 *Assessment and Data Collection*

Projects developed by pre-service ECTs were evaluated using an instructor-developed rubric assessing four key dimensions: pedagogical alignment with early childhood learning objectives, integration of scientific content, effective use of ScratchJr's technical features, and overall presentation quality.

To assess self-efficacy and technology acceptance, pre-service ECTs completed a post-course questionnaire. The instrument included 28 Likert-scale items rated on a 5-point scale (1 = strongly disagree to 5 = strongly agree). Seven items were adapted from the Teachers' Self-Efficacy in Computational Thinking (TSECT) instrument [24], focusing on pre-service ECTs' perceived confidence in their ability to teach CT and introductory coding skills. The remaining 21 items were based on Davis's Technology Acceptance Model (TAM) [25], adapted to evaluate pre-service ECTs' perceptions of ScratchJr. TAM has been widely used in educational technology contexts, including ER and CT; for example, it has been used to analyze teachers' responses to open-ended questions to identify and determine their views regarding the perceived usefulness and perceived ease-of-use of floor-robots as a classroom technological tool [8, 26]. It has also been used to examine Computer Science teachers' perceptions, beliefs, and attitudes on Computational Thinking [27].

The TAM framework consists of four core constructs: Perceived Ease of Use (PEOU)—defined as "the degree to which a pre-service ECT believes that using ScratchJr would be free of effort"; Perceived Usefulness (PU)—defined as "the degree to which a pre-service ECT believes that using ScratchJr would enhance their professional performance"; Attitude Toward Use (ATT)—referring to "the overall affective evaluation of using ScratchJr"; and Behavioral Intention (BI)—as "the pre-service ECTs' intention or likelihood to use ScratchJr in the future".

For the purposes of the present study, an additional external variable was incorporated: Anxiety (ANX)—defined as "feelings of apprehension or discomfort related to the use of ScratchJr". This construct aimed to investigate how anxiety may shape pre-service ECTs' perceptions and attitudes toward adopting ScratchJr. The rationale for its inclusion stems from the premise that pre-service ECTs who experience anxiety regarding technology are more likely to perceive it as difficult to use [28]. Furthermore, elevated levels of anxiety can directly undermine the intention to use technological tools such as ScratchJr, even when their usefulness is recognized. It is our belief that embedding CT practices within pre-service ECT education programs can play a crucial role in reducing such anxieties and fostering greater confidence in the integration of digital tools in ECE.

The questionnaire was translated, culturally adapted, and validated for the Greek context. A pilot study was conducted to confirm the clarity, reliability, and internal consistency of the instrument. Ethical procedures were strictly followed, including obtaining informed consent from all participants, ensuring compliance with the General Data Protection Regulation (GDPR), and maintaining participant anonymity and data confidentiality. Quantitative data from the questionnaire were analyzed using SPSS 29.0 for Windows.

3 Results and Discussion

3.1 Project Themes and Formats

According to the first research question, the ScratchJr projects developed by the pre-service early childhood ECTs were categorized into three major science domains, consistent with Greece's national ECE curriculum: "Matter and Phenomena", "Living Organisms", and "Earth – Planetary System and Space".

The "Matter and Phenomena" category attracted the highest number of projects (20 out of 45 projects), with a strong emphasis on the Water Cycle (13 out of 20 projects). This indicates that pre-service ECTs gravitate toward topics that are visually dynamic and lend themselves well to storytelling and animation. As Bers [29] explains, ScratchJr is designed to support young children's learning by providing a visual programming environment that facilitates narrative expression and creative exploration. This supports the idea that topics which are concrete and observable are more easily translated into digital formats, a finding consistent with Eshach [30], who emphasized the importance of tangibility in ECSE. Conversely, the low representation of abstract subtopics such as Friction, Changes in Matter, and Recycling (1 out of 20 projects) suggests challenges in conveying conceptually complex ideas within the constraints of a child-friendly programming tool, a phenomenon also highlighted by Fleer [31].

In the "Living Organisms" domain (11 out of 45 projects), the overwhelming focus on the basic characteristics of living beings (10 out of 11 projects) reflects a tendency to select general and familiar biological concepts, likely due to their immediate relevance and recognizability to young learners. This aligns with Robin [32], who found that digital storytelling in educational settings often centers around themes that are intuitively meaningful to both the creator and the audience. The limited focus on ecological subtopics such as habitats (1 out of 11 projects) points to a need for further support and scaffolding when approaching more complex environmental themes.

The projects in the "Earth – Planetary System and Space" category (14 out of 45 projects) exhibited a more balanced thematic distribution, particularly favoring cyclical natural phenomena like day-night and seasonal cycles (6 out of 14 projects). These topics are inherently suitable for animation and narrative sequencing, making them accessible even within the technical limitations of beginner programming platforms. This reflects the observations of Strawhacker and Bers [17], who noted that ScratchJr is particularly effective when used to represent patterns and repetitive sequences, a key feature of many natural cycles taught in ECSE.

With regard to project format, the clear preference for tutorial-type activities (26 out of 45 projects) underscores a pedagogical inclination toward structured and guided learning. This may reflect the pre-service ECTs' desire to create clear, goal-oriented instructional content, as also reported by Angeli and Valanides [33], who highlighted the link between pre-service ECTs' pedagogical intentions and their integration of technology. However, the use of storytelling (10 out of 45 projects) and game-based formats (9 out of 45 projects) reveals an appreciation for interactive and narrative-driven teaching strategies, consistent with the work of Ohler [34] on digital storytelling as a powerful pedagogical tool in early education.

Finally, the analysis of programming skills demonstrated varied levels of proficiency among the pre-service ECTs. While a notable portion of projects (18/45) showcased a high level of ScratchJr command usage, indicating creative and confident use of programming blocks, many projects reflected intermediate to medium levels of competence, with only a few at the low end. This is consistent with findings by Tondeur et al. [35], who emphasized that technological confidence and pedagogical beliefs play a critical role in shaping the way teachers adopt and use digital tools in classroom settings. The generally moderate-to-high performance levels suggest that the pre-service ECTs were able to engage with ScratchJr meaningfully, even if not all achieved full mastery.

Figure 1 illustrates a project characterized by an advanced use of Scratch commands, focused on the "Matter and Phenomena" category, with particular emphasis on the concept of Magnetism.

3.2 Engagement and Learning Outcomes

Observations and student reflections suggested that the integration of ScratchJr enhanced pre-service ECT engagement with both science content and pedagogical design. During classes many pre-service ECTs reported that the process of breaking down a scientific concept into a simple, visual story helped them clarify

Fig. 1 Exploring magnetism through ScratchJr: a project on matter and phenomena

their own understanding of the material. The creative freedom afforded by ScratchJr was frequently highlighted as a motivating factor that enhanced both the accessibility and the enjoyment of engaging with science content. Despite initial apprehension about programming, pre-service ECTs quickly became comfortable with the ScratchJr interface. Peer support and collaborative exploration played a key role in helping pre-service ECTs overcome early technical challenges.

3.3 Questionnaires' Validity

The research model presents a high level of reliability and consistency as Cronbach's Alpha exceeds the recommended value of 0.7 ($a = 0.912$, Number of items = 28) well above DeVellis's [36] recommendations (>0.70). Their validity was also measured and included in the measurement of each of the constructs that were proposed as shown in Table 1.

Table 1 Validity of research's' questionnaire

Factor	No of questions	Cronbach's alpha
TSECT	7	0.865
PU	4	0.883
PEOU	4	0.879
ATT	4	0.905
ANX	4	0.740
BI	5	0.908

3.4 Pre-service Early Childhood Teachers' Self-Efficacy in ScratchJr

Regarding the second research question, the construct of Teachers' Self-Efficacy in Computational Thinking (TSECT) was used in order to capture pre-service ECTs' perceived ability to incorporate programming and CT into future science teaching practices. The questionnaire analysis revealed a positive shift in pre-service ECTs' confidence in teaching science using ScratchJr (Cronbach's alpha $\alpha = 0.865, N = 7$). Pre-service ECTs responded using a five-point Likert scale (1 = not at all to 5 = to a great extent). For each statement, the mean (M), standard deviation (SD), and mode (Mo) are presented (see Table 2). The results indicate that pre-service ECTs felt a high degree of confidence in creating simple projects with ScratchJr ($M = 4.49$, SD = ± 0.84), which reflects a solid familiarity with the platform and its core functionalities. They also demonstrated a satisfactory level of instructional self-efficacy, albeit at a slightly lower level ($M = 3.87$, SD = ± 0.87) compared to their technical use of the tool. Furthermore, they expressed strong trust in the pedagogical potential of ScratchJr to foster positive attitudes toward visual programming and CT ($M = 4.22$, SD = ± 0.85). However, significant variation was observed in pre-service ECTs' self-perception regarding their role as facilitators in guiding the creation of ScratchJr projects ($M = 3.76$, SD = ± 1.31). Pre-service ECTs also exhibited reservations about the implementation of CT within the ECE curriculum ($M = 3.73$, SD = ± 0.89), as well as about engaging in collaborative efforts and disseminating educational material ($M = 3.62$, SD = ± 0.96). Finally, a notable degree of uncertainty or perceived inadequacy was expressed regarding the design of complex instructional scenarios that integrate CT ($M = 3.51$, SD = ± 1.01).

3.5 Pre-Service Early Childhood Teachers' Acceptance of ScratchJr for Learning and Teaching Computational Thinking in Science Education

Regarding the third research question, the results are presented below for each construct measured by the questionnaire.

Table 2 Descriptive statistics for pre-service ECTs' self-efficacy in computational thinking

Question	$M \pm$ SD	Mo
I feel confident in creating simple projects with ScratchJr	4.49 ± 0.84	5
I know how to teach programming concepts with ScratchJr	3.87 ± 0.87	4
I can encourage a positive attitude toward programming with ScratchJr to my students	4.22 ± 0.85	5
I can act as a mentor teacher and support other teachers or students to use CT and coding as an interdisciplinary tool	3.76 ± 1.31	4
I am sure to implement CT concepts and coding as an educative tool within the preschool classroom	3.73 ± 0.89	3
I can share lesson plans, activity sheets based on CT, and coding as an educative tool within the preschool classroom	3.62 ± 0.96	3
I can create lesson plans using CT, algorithmic, and coding as an educative tool	3.51 ± 1.01	3

Table 3 Descriptive statistics of pre-service ECTs' perceptions on the usefulness toward ScratchJr

Question	$M \pm$ SD	Mo
Using ScratchJr will improve the performance of my teaching.	4.49 ± 0.84	4
Using ScratchJr will enhance the effectiveness of my teaching.	4.22 ± 0.67	4
Using ScratchJr will increase my productivity in teaching.	3.96 ± 0.77	4
ScratchJr will be useful for my teaching.	4.24 ± 0.71	3

3.5.1 Evaluation of Pre-service Early Childhood Teachers' Perceptions on the Usefulness Toward ScratchJr

This section evaluates pre-service ECTs' views regarding the Perceived Usefulness (PU) of ScratchJr as a tool for enhancing the teaching and learning process. According to Table 3, Perceived Usefulness construct indicate that pre-service ECTs consider ScratchJr to be a highly valuable tool ($M = 3.96$, SD = ± 0.64), capable of enhancing teaching effectiveness ($M = 4.22$, SD = ± 0.67), as well as improving instructional performance and productivity ($M = 3.96$, SD = ± 0.77). The high mean values and low standard deviations reinforce the image of a widely accepted tool, highlighting its strong potential for integration into the ECE curriculum ($M = 4.24$, SD = ± 0.71).

3.5.2 Evaluation of Pre-service Early Childhood Teachers' Perceptions on the Ease of Use Toward ScratchJr

The results for the Perceived Ease of Use (PEOU) factor suggest that pre-service ECTs generally perceive ScratchJr as a user-friendly ($M = 4.11$, SD = ± 0.80) and interactively accessible tool ($M = 4.09$, SD = ± 0.95), suitable for integration into

Table 4 Descriptive statistics of pre-service ECTs' perceptions on the ease of use toward ScratchJr

Question	$M \pm$ SD	Mo
It is easy for me to learn to use ScratchJr for my teaching	4.11 ± 0.80	5
My interaction with ScratchJr is clear and understandable	4.09 ± 0.95	5
It is easy for me to find the ScratchJr app in order to integrate it into my teaching	4.07 ± 0.89	5
In general, I find it easy to use ScratchJ in my teaching	4.09 ± 0.87	4

preschool teaching. The high mean scores and Mode values at the upper end of the scale reinforce the positive attitude toward its use, while the relatively low standard deviations indicate a degree of homogeneity in participants' perceptions. Overall, the ease of access ($M = 4.07$, SD = ± 0.89), comprehension, and use ($M = 4.09$, SD = ± 0.87) of the application is considered a significant factor in technology acceptance among the pre-service ECTs. The majority of the trainees strongly agree with the positive value of ScratchJr in teaching practice, a finding that is reflected in both the mean and the Mode values (Table 4).

3.5.3 Evaluation of Pre-service Early Childhood Teachers' Perceptions on Attitude Toward ScratchJr

The results for the Attitudes (ATT) factor (see Table 5) indicate that pre-service ECTs hold a very positive attitude toward the use of ScratchJr in the educational process ($M = 4.40$, SD = ± 0.78). They report that the use of ScratchJr significantly contributes to engaging pre-service ECTs' interest during the teaching process ($M = 4.53$, SD = ± 0.66), while also expressing a positive stance regarding the emotional dimension of using the tool, indicating a predisposition for enjoyment and creativity in its use ($M = 4.38$, SD = ± 0.72). Finally, pre-service ECTs more broadly recognize the value of integrating ScratchJr into the curriculum of preschool or ECE ($M = 4.33$, SD = ± 0.80).

Table 5 Descriptive statistics of pre-service ECTs' perceptions on attitudes toward ScratchJr

Question	M ± SD	Mo
Using ScratchJr in teaching is a good idea	4.40 ± 0.78	5
Using ScratchJr makes teaching more engaging	4.53 ± 0.66	5
I feel that my work with ScratchJr will be enjoyable	4.38 ± 0.72	5
In my opinion, it would be desirable to use ScratchJr in education	4.33 ± 0.80	5

Table 6 Descriptive statistics of pre-service ECTs' anxiety toward ScratchJr

Question	M ± SD	Mo
I feel scared/afraid about using ScratchJr	1.87 ± 1.04	1
It scares me to think that I might lose valuable information by pressing a button incorrectly when using ScratchJr	1.96 ± 1.21	1
I hesitate to use ScratchJr for fear of making a mistake that I can't fix	1.47 ± 0.66	1
ScratchJr somehow scares me into using it	1.58 ± 0.69	1

3.5.4 Evaluation of Pre-service Early Childhood Teachers' Anxiety Toward ScratchJr

The findings indicate very low levels of anxiety and fear (M = 1.87, SD = ± 1.04) toward the use of ScratchJr by pre-service ECTs. According to the results from Table 6, they report feeling comfortable and secure when navigating (M = 1.96, SD = ± 1.21) and experimenting (M = 1.47, SD = ± 0.66) with the tool, which positively contributes to its overall acceptance and integration into educational practice (M = 1.96, SD = ± 1.21). The fact that the dominant response to all statements is the value "1—not at all" further reinforces the observation that anxiety does not hinder the integration of technology into ECE.

3.5.5 Evaluation of Pre-service Early Childhood Teachers' Perceptions on the Intention to Use ScratchJr

With regard to the Behavioral Intention (BI) factor, the results from Table 7 reveal a positive intention among pre-service ECTs to adopt and promote the use of ScratchJr (M = 3.98, SD = ± 0.89). Despite some variation in the frequency of intended use, there is a clear trend toward integrating ScratchJr into their pedagogical repertoire (M = 3.53, SD = ± 1.01), a finding also supported by the mode value of 4, indicating that pre-service ECTs consider the tool suitable for regular use, albeit with some variability. However, they appear to have some reservations (M = 3.49, SD = ± 1.06) regarding frequent application of the tool in their teaching practice, as suggested by the relatively high standard deviation. On the other hand, they express a strong willingness to share their positive experience of using ScratchJr within their broader professional and social networks (M = 4.04, SD = ± 0.74). Finally, pre-service ECTs report a strong intention to support and recommend ScratchJr as an educational tool within the educational community (M = 4.22, SD = ± 0.74).

Table 7 Descriptive statistics of pre-service ECTs' perceptions on the intension to use ScratchJr

Question	M ± SD	Mo
I will use ScratchJr in my teaching	3.98 ± 0.89	4
I plan to use ScratchJr frequently in my teaching	3.53 ± 1.01	4
I plan to use ScratchJr as often as possible in my teaching	3.49 ± 1.06	3
I will refer to the positive perspectives of using ScratchJr in education, to my friends and fellow students	4.04 ± 0.74	4
I will recommend the use of ScratchJr to my future colleagues	4.22 ± 0.74	4

4 Conclusion

This study highlights the significant potential of integrating visual programming and, in particular ScratchJr, into ECSE within the context of pre-service ECTs training. The findings reveal critical trends in how pre-service ECTs engage with digital technologies, particularly ScratchJr, to develop science-oriented instructional activities. These results are interpreted in light of existing literature, offering both theoretical and empirical support.

A notable preference for tutorial-based project formats reflects a pedagogical tendency toward structured and goal-oriented learning, consistent with prior findings [33]. Simultaneously, the presence of storytelling and game-based formats indicates a recognition of interactive and narrative-driven teaching strategies, echoing Ohler's [34] advocacy for digital storytelling in ECE.

Pre-service ECTs' programming proficiency varied, with 18 projects demonstrating a high level of ScratchJr command, suggesting confident and creative use of the platform. Most projects exhibited moderate competence, supporting prior research [35] that highlights the role of technological self-efficacy and pedagogical beliefs in the effective adoption of digital tools. Despite limited prior experience with programming, pre-service ECTs displayed a significant increase in confidence regarding both science instruction and technology integration, addressing known gaps in ECSE preparation [1, 4].

Engagement with developmentally appropriate visual programming environments, such as ScratchJr, [11, 12] enabled pre-service ECTs to design curriculum-aligned science projects that enhanced both content knowledge and digital confidence. These findings align with literature advocating for the promotion of CT through accessible, child-friendly platforms [15, 16].

Importantly, the study aligns with the Technology Acceptance Model, with pre-service ECTs reporting high levels of perceived usefulness and ease of use—factors that positively influenced their behavioral intention to implement ScratchJr in future classrooms [27]. Furthermore, the results support existing evidence that visual programming tools foster creativity, engagement, and higher-order thinking through inquiry-based learning [17, 18], thus positioning ScratchJr as both a pedagogical and technological bridge.

The findings emphasize the necessity of embedding visual programming environments into ECTs education programs to develop technologically competent ECTs who can foster scientific inquiry from an early age [8, 37, 38]. While pre-service ECTs expressed strong intentions to integrate ScratchJr into their pedagogical practices, the study also highlights areas for further growth, such as deeper conceptual engagement and expansion of thematic coverage.

Limitations of the study must be acknowledged. One key limitation is the relatively small sample size, which may constrain the generalizability of the findings to broader populations of pre-service ECTs. Furthermore, a regression analysis was conducted to examine the relationships among the core constructs of the research model and the Behavioral Intention of pre-service ECTs to utilize ScratchJr in their future teaching practice. While the analysis offers valuable insights, the scope of interpretation remains context-specific and should be approached with caution. Moreover, the findings are situated within a specific institutional and curricular context in Greece, potentially limiting transferability. Future research should explore replicating this model in diverse educational settings and investigate the long-term pedagogical impacts of ScratchJr, particularly with respect to interdisciplinary curriculum design and the cultivation of professional learning communities to support sustainable technology integration in ECE.

References

1. Cabe Trundle, K.: The inclusion of science in early childhood classrooms. In: Research in Early Childhood Science Education, pp. 1–6. Springer, Dordrecht (2015)
2. Ravanis, K.: Research trends and development perspectives in early childhood science education: an overview. Educ. Sci. **12**, 456 (2022). https://doi.org/10.3390/educsci12070456
3. Cordova Jr., N., Kilag, O.K., Andrin, G., Groenewald, E., Abella, J.: Promoting literacy in early childhood: leadership practices and long-term educational impact. Excellencia: international multi-disciplinary. J. Educ. **2**(1), 79–89 (2024)
4. Balabani, A., Lavidas, K., Meli, K., Koliopoulos, D.: Attitudes and beliefs of Greek preschool teachers on the teaching of science. Rev. Sci. Math. ICT Educ. **18**(2), 89–107 (2024). https://doi.org/10.26220/rev.5238
5. Sullivan, A., Heffernan, J.: Robotics in the early childhood classroom: learning outcomes from an 8-week robotics curriculum in pre-kindergarten through second grade. Int. J. Technol. Des. Educ. **26**, 3–20 (2016). https://doi.org/10.1007/s10798-015-9304-5
6. Tselegkaridis, S., Sapounidis, T.: A systematic literature review on STEM research in early childhood. In: Papadakis, S., Kalogiannakis, M. (eds.) STEM, Robotics, Mobile Apps in Early Childhood and Primary Education Lecture Notes in Educational Technology, pp. 117–134. Springer, Singapore (2022). https://doi.org/10.1007/978-981-19-0568-1_7
7. Alimisis, D.: Educational robotics: open questions and new challenges. Themes Sci. Technol. Educ. **6**(1), 63–71 (2013)
8. Schina, D., Valls-Bautista, C., Borrull-Riera, A., Usart, M., Esteve-González, V.: An associational study: preschool teachers' acceptance and self-efficacy towards educational robotics in a pre-service teacher training program. Int. J. Educ. Technol. High. Educ. **18**, 1–20 (2021). https://doi.org/10.1186/s41239-021-00264-z
9. Wing, J.M.: Computational thinking. Commun. ACM. **49**(3), 33–35 (2006)
10. Bers, M.U., Strawhacker, A., Sullivan, A.: The State of the Field of Computational Thinking in Early Childhood Education. OECD, Paris (2022). https://doi.org/10.1787/3354387a-en

11. Angeli, C., Valanides, N.: Developing young children's computational thinking with educational robotics: a study on problem-solving and programming capabilities in early childhood. Early Childhood Educ. J. **105**, 105954 (2020). https://doi.org/10.1016/j.chb.2019.03.018
12. Bers, M.U.: The tangibleK robotics program: applied computational thinking for young children. Early Childhood Res. Pract. **12**(2), 2 (2010)
13. Chaldi, M., Mantzanidou, I.: Educational robotics and STEAM in early childhood education. Adv. Mobile Learn. Educ. Res. **1**(2), 72–81 (2021). https://doi.org/10.25082/AMLER.2021.02.003
14. Muñoz-Repiso, A.G.V., Caballero-González, Y.A.: Robotics to develop computational thinking in early childhood education. Comun. Media Educ. Res. J. **27**(59), 63–72 (2019). https://doi.org/10.3916/C59-2019-06
15. Louka, A., Papadakis, S.: Enhancing computational thinking in early childhood education through ScratchJr integration. Heliyon. **10**(10), e30482 (2024). https://doi.org/10.1016/j.heliyon.2024.e30482
16. Strawhacker, A., Lee, M., Bers, M.U.: Teaching tools, teachers' rules: exploring the impact of teaching styles on young children's programming knowledge in ScratchJr classrooms. Int. J. Technol. Des. Educ. **28**, 347–376 (2018). https://doi.org/10.1007/s10798-017-9400-9
17. Strawhacker, A., Bers, M.U.: What they learn when they learn coding: investigating cognitive domains and computer programming knowledge in young children. Educ. Technol. Res. Dev. **67**, 541–575 (2019). https://doi.org/10.1007/s11423-018-9622-x
18. Papadakis, S., Kalogiannakis, M., Zaranis, N.: Developing fundamental programming concepts and computational thinking with ScratchJr in preschool education: a case study. Int. J. Mobile Learn. Org. **10**(3), 187–202 (2017). https://doi.org/10.1504/IJMLO.2016.077867
19. Bers, M.U.: Coding and computational thinking in early childhood: the impact of ScratchJr in Europe. Eur. J. STEM Educ. **3**(3), 3868 (2018). https://doi.org/10.20897/ejsteme/3868
20. Portelance, D.J., Strawhacker, A., Bers, M.U.: Constructing the ScratchJr programming language in the early childhood classroom. Int. J. Technol. Des. Educ. **26**, 489–504 (2016). https://doi.org/10.1007/s10798-015-9325-0
21. Flannery, L.P., Bers, M.U., Kazakoff, E.R., Sullivan, A.: Designing ScratchJr: Support for early childhood learning through computer programming. In: Proceedings of the 12th International Conference on Interaction Design and Children, pp. 1–10. ACM, New York (2013). https://doi.org/10.1145/2485760.2485785
22. Louka, A.: Programming environments for the development of CT in preschool education: a systematic literature review. Adv. Mobile Learn. Educ. Res. **3**(1), 525–540 (2022). https://doi.org/10.25082/AMLER.2023.01.001
23. Kalogiannakis, M., Papadakis, S., Zaranis, N.: Developing fundamental programming concepts and computational thinking with ScratchJr in preschool education: a case study. Int. J. Mobile Learn. Org. **10**(3), 187–202 (2016). https://doi.org/10.1504/IJMLO.2016.0778
24. Bean, N., Weese, J., Feldhausen, R., Bell, R.: Starting from scratch: Developing a pre-service teacher training program in computational thinking. In: 2015 IEEE Frontiers in education conference (FIE), pp. 1–8. IEEE, New York (2015). https://doi.org/10.1109/FIE.2015.7344237
25. Davis, F.D.: Perceived usefulness, perceived ease of use, and user acceptance of information technology. MIS Q. **13**(3), 319–340 (1989). https://doi.org/10.2307/249008
26. Casey, J.E., Pennington, L.K., Mireles, S.V.: Technology acceptance model: assessing preservice teachers' acceptance of floor-robots as a useful pedagogical tool. Technol. Knowl. Learn. **26**(3), 499–514 (2021). https://doi.org/10.1007/s10758-020-09452-8
27. Fessakis, G., Prantsoudi, S.: Computer science teachers' perceptions, beliefs and attitudes on computational thinking in Greece. Inf. Educ. **18**(2), 227–258 (2019). https://doi.org/10.15388/infedu.2019.11
28. Tsai, T.H., Lin, W.Y., Chang, Y.S., Chang, P.C., Lee, M.Y.: Technology anxiety and resistance to change behavioral study of a wearable cardiac warming system using an extended TAM for older adults. PLoS One. **15**(1), e0227270 (2020). https://doi.org/10.1371/journal.pone.0227270

29. Bers, M.U.: Coding as a Playground: Programming and Computational Thinking in the Early Childhood Classroom. Routledge, London (2020). https://doi.org/10.4324/9781003022602
30. Eshach, H.: Science Literacy in Primary Schools and Pre-Schools: A Guide for Teachers. Springer, Dordrecht (2006)
31. Fleer, M.: Supporting scientific conceptual consciousness or learning in 'a roundabout way' in play-based contexts. Int. J. Sci. Educ. **31**(8), 1069–1089 (2009). https://doi.org/10.1080/09500690801953161
32. Robin, B.R.: Digital storytelling: a powerful technology tool for the 21st century classroom. Theory Pract. **47**(3), 220–228 (2008). https://doi.org/10.1080/00405840802153916
33. Angeli, C., Valanides, N.: Preparing preservice elementary teachers to teach science through computer models. Contemp. Issues Technol. Teacher Educ. **6**(1), 87–98 (2006)
34. Ohler, J.: Digital Storytelling in the Classroom: New Media Pathways to Literacy, Learning, and Creativity, 2nd edn. Corwin Press, Thousand Oaks (2013). https://doi.org/10.4135/9781452277479
35. Tondeur, J., Scherer, R., Siddiq, F., Baran, E.: A comprehensive investigation of TPACK within pre-service teachers' ICT profiles: mind the gap! Australas. J. Educ. Technol. **33**(3), 3504 (2017). https://doi.org/10.14742/ajet.3504
36. DeVellis, R.F., Thorpe, C.T.: Scale Development: Theory and Applications. Sage Publications, London (2021)
37. Dong, W., Li, Y., Sun, L., Liu, Y.: Developing pre-service teachers' computational thinking: a systematic literature review. Int. J. Technol. Des. Educ. **34**(1), 191–227 (2024). https://doi.org/10.1007/s10798-023-09811-3
38. Kalogiannakis, M., Papadakis, S.: Evaluating pre-service kindergarten teachers' intention to adopt and use tablets into teaching practice for natural sciences. Int. J. Mobile Learn. Org. **13**(1), 113–127 (2018). https://doi.org/10.1504/IJMLO.2019.096479

Investigating Greek Teachers' Training Needs in Digital Competence: Validation of the DigCompEdu Check-in Scale Using Structural Equation Modeling

George Chorozidis and Charalampos Karagiannidis

Abstract The impact of digital transformation on various aspects of daily life has necessitated the demand for digitally competent teachers who can facilitate students' digital competence in a world increasingly mediated by digital technologies. This study is based on the European Framework for the Digital Competence of Educators (DigCompEdu), which is considered a comprehensive framework for describing the multifaceted nature of teachers' digital competence. The study aims to assess the reliability and validity of the Greek version of the DigCompEdu Check-In as a self-reflection instrument and to determine the participants' level of digital competence, both overall and by competence areas, in accordance with the DigCompEdu framework. This empirical research followed a quantitative, non-experimental, descriptive, correlational design and involved 208 in-service primary and secondary teachers in Greece. Exploratory and Confirmatory Factor Analysis (EFA and CFA) were performed using structural equation modeling (SEM) to determine the factorial structure of the instrument. The analysis confirmed the reliability and validity of the scale, as well as its underlying factorial structure. The findings indicated that the overall proficiency level of the participating teachers is moderate and identified "Teaching and Learning," "Assessment," and "Facilitating learners' digital competence" as the least developed areas. These results underscore the need for teachers to increase their digital competence in the aforementioned areas through specific training to effectively integrate digital technology into their teaching practice.

Keywords Digital competence · DigCompEdu · Self-perception · Training needs · Structural equation modeling

G. Chorozidis (✉) · C. Karagiannidis
Department of Special Education, University of Thessaly, Volos, Greece
e-mail: gchorozidis@uth.gr

C. Karagiannidis
e-mail: karagian@uth.gr

I. Kazanidis and A. Tsinakos (eds.), *Research on E-Learning and ICT in Education*,
https://doi.org/10.1007/978-3-031-99961-1_5

1 Introduction

Digital technologies have assumed a fundamental role at every level of the so-called knowledge society and have significantly changed how people learn, communicate, work, and interact with each other. Therefore, it is vital to recognize and adapt to these changes to fully capitalize on the potential offered by digital technologies and overcome the challenges arising from their ever-increasing use in our daily lives [1]. Digital competence is one of the eight key competences for lifelong learning and entails a set of skills, knowledge, attitudes and strategies for the confident, critical and responsible use of digital technologies to perform tasks in real-life situations [2–4].

To this extent, digital competence is central to addressing the challenges posed by the proliferation of digital technologies, and education is considered the key to supporting its development [5]. The need to equip citizens with this competence to participate in civic, social and professional life, places new demands and expectations upon teachers at all levels of education [6]. Teachers, therefore, must be digitally competent and leverage the power of digital technologies in their teaching practices to effectively promote students' digital competence and prepare them for the demands of the digital age [7, 8]. This has led to an increased focus on teachers' digital competence. However, what exactly constitutes teachers' digital competence and its related factors still remain a topic of debate [9].

In this respect, teachers' digital competence is considered an additional layer of their teaching profession, a fact that is recognized by many education-focused policies [10]. Consequently, there is a growing concern about defining the concept of digital competence and addressing its multifaceted nature. In response, numerous national and international frameworks have been developed to capture this complexity, serving as a basis for describing the specific competences teachers need to become digitally competent [11]. A comprehensive evaluation of the most commonly used digital competence frameworks in international contexts through expert judgment is presented in [12]. Among these, the European Framework for the Digital Competence of Educators (DigCompEdu) is the most highly valued framework compared to others.

The European Framework for the Digital Competence of Educators (DigCompEdu) was published at the end of 2017 by the Joint Research Center of the European Union as the result of numerous scientific studies con-ducted at local, national, and international levels. It provides a common language and logic to facilitate discussions and exchange of best practices across borders. The DigCompEdu framework explicitly describes what it means for educators to be digitally competent and is suitable for educators at all levels of education, including vocational and adult education. This framework is designed to serve as a common reference point, synthesizing existing models into a single, generic, transnational framework that recognizes the need for flexibility and adaptability across diverse social settings. The main advantage of the DigCompEdu framework is that it emphasizes the pedagogical dimension of digital technologies by providing specific guidance for teachers' pedagogical practices [13, 14].

The DigCompEdu framework proposes 22 educator-specific competences organized in 6 competence areas, as Fig. 1 shows. The core of the DigCompEdu framework focuses on educators' pedagogical competences, as defined by areas 2–5, which outline the digital competences educators need to develop an efficient, inclusive, and innovative learning environment. Area 2 (Digital Resources) proposes the competences that teachers need to select, create, distribute, and protect digital resources to support their teaching. Area 3 (Teaching and Learning) entails competences to plan, design, and implement digital technologies in different stages of the teaching and learning process. Area 4 (Assessment) deals with the use of digital technologies to shift from traditional assessment methods to facilitate innovative assessment strategies and provide targeted feedback using digital technologies. Area 5 (Empowering Learners) prioritizes student-centered learning and emphasizes the potential of digital technologies to boost the active participation in the learning process and their ownership of it. The remaining competence areas specify the educators' professional competences and learner-oriented competences. Area 1 (Professional Engagement) focuses on the utilization of digital technologies for organizational communication, professional collaboration, and continuous professional development. Area 6 (Facilitating Learners' Digital Competence) encompasses an integral part of teachers' digital competence, which is the ability of teachers to enable students to use digital technologies creatively and responsibly [8, 13].

The DigCompEdu framework, similarly to other frameworks, also proposes a progression model of digital competence. It contains six progressively advanced proficiency levels, each accompanied by a self-assessment descriptor. The proposed progression model aligns with the six proficiency levels used by the Common European Framework of Reference for Languages (CEFR), ranging from Newcomer (A1) to Pioneer (C2). In this regard, the initial two levels, Newcomer (A1) and Explorer (A2), teachers focus on acquiring new information and developing basic digital practices. At the intermediate levels, Integrator (B1) and Expert (B2), teachers apply the knowledge they have gained while seeking ways to expand and reflect on their digital teaching practices. Finally, at the highest levels, Leader (C1) and Pioneer (C2), teachers apply their knowledge and share it with others, critically analyze existing practices, and actively develop new approaches to their teaching profession [13]. To support this process of self-assessment and professional growth, a self-assessment instrument called DigCompEdu Check-In, was developed in collaboration with a multi-national team of researchers, under the auspices of the European Commission Joint Research Centre [8].

Despite the growing interest in teachers' digital competence and the availability of several competence frameworks, there is a lack of validated instruments adapted to the Greek context. The aim of this study is to validate the Greek version of the DigCompEdu Check-In self-assessment questionnaire for primary and

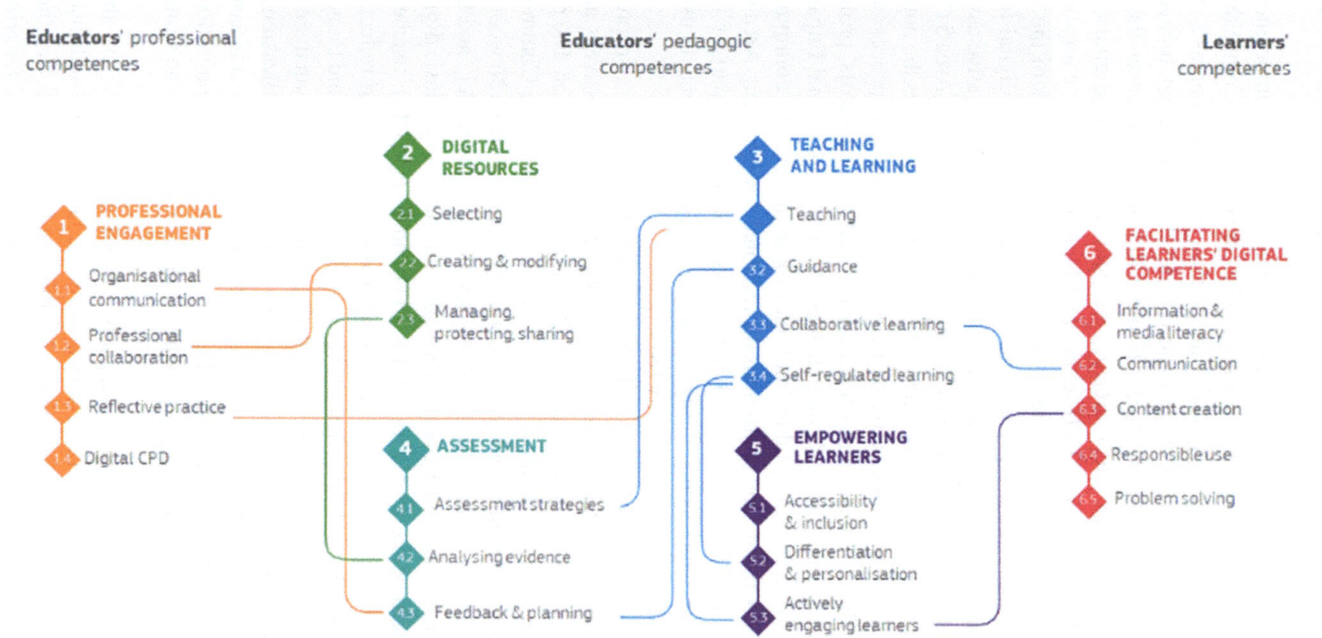

Fig. 1 The DigCompEdu framework in a nutshell [13]

secondary education teachers and to identify their training needs in digital competence according to the six competence areas proposed by the DigCompEdu framework. Specifically, the first objective of this research is to measure the reliability and validity of the Greek version of this instrument. The second objective is to examine the training (educational) needs of a sample of Greek teachers in terms of their digital competence, overall and by areas of competence, according to the DigCompEdu framework, to identify their strengths and areas of improvement.

In the context of this research, a training (educational) need is defined as any knowledge, skill, or attitude regarding the use of digital technologies in teaching practice that is essential but currently unavailable to teachers. The investigation of training needs was conducted based on the structural-functional and psycho-sociological approach to the concept of need, emphasizing both the element of change—stemming from the constant and rapid development of digital technologies in today's era—and the element of deficiency, reflecting the reality that teachers' knowledge and skills are often in-sufficient to keep pace with technological advancements [15].

2 Materials and Methods

The empirical component of this research follows a quantitative methodological approach with a non-experimental descriptive correlational design, as it allows for the investigation of relationships among variables [16]. This cross-sectional research study intends to evaluate the reliability and validity evidence of the Greek version of the DigCompEdu Check-In self-assessment questionnaire. For this, Exploratory Factor Analysis (EFA) and Confirmatory Factor Analysis (CFA) were conducted using structural equation modeling (SEM). Furthermore, in alignment with the research aim, this methodology enables the collection of information on teachers' self-perceived level of digital competence and the identification of their training needs in terms of digital competence, as outlined in the six areas that constitute the DigCompEdu framework.

2.1 Sample

The study sample consisted of 208 primary and secondary education teachers in Greece, selected using a non-probabilistic purposive sampling method. Of the respondents, 77% were women and 23% were men. More specifically, 45% of the teachers work in secondary education, while 55% work in primary education. Of those in primary education, 60% are employed in special education, and the remaining 40% in general education. Among secondary school teachers, 55% specialize in the humanities and social sciences, while 45% specialize in natural sciences, including mathematics, physics, and chemistry.

2.2 Data Collection Instrument

The instrument used for data collection was the DigCompEdu Check-In self-assessment questionnaire, in a version translated and adapted for use in the Greek context. This self-assessment tool is designed to help teachers self-assess their perceived level of digital competence according to the key concepts of the DigCompEdu framework. The instrument consists of 22 statements, each corresponding to one of the 22 competences distributed across the six competence areas outlined by the DigCompEdu framework. Each statement is followed by a Likert-type scale with five response intervals, where respondents indicate the extent to which the statement reflects their own practice by selecting one of five response options. These response options are organized progressively, reflecting the overall progression logic of the DigCompEdu framework through an internal scoring system. Each response is scored from 0 to 4 and the maximum total score is 88 points, which corresponds to the six proficiency levels of competence (A1 to C2). Table 1 presents the scoring rule of the DigCompEdu self-assessment instrument, overall and by areas of competence, according to the six progressive proficiency levels proposed by the DigCompEdu framework.

The DigCompEdu Check-In self-assessment questionnaire goes beyond merely listing a set of competences by providing respondents with a detailed feedback report upon submission to help them identify their professional development needs. Based on the answers given, the report indicates the score per competence area and offers detailed feedback, including suggestions for improving current teacher practices. This detailed feedback helps teachers recognize their strengths and focus on less developed areas to further enhance their digital competence. In this regard, the added value of this self-assessment tool lies in enabling respondents not only to determine their current perceived level of digital competence but also to receive specific suggestions for improvement. Through this process of self-assessment and professional growth, teachers are guided on the next steps to take to move beyond to the next level and enhance their proficiency level in the least developed areas.

It is worth mentioning that the DigCompEdu Check-In self-assessment questionnaire has been validated across various European countries. To this end, this article contributes to the validation of this questionnaire by ensuring that its translation

Table 1 Scoring rule and classification system for DigCompEdu Check-In self-assessment questionnaire, overall and by areas of competence

Areas of competence	Items	Scores	DigCompEdu progression levels					
			A1	A2	B1	B2	C1	C2
Total	22	0–88	0–19	20–33	34–49	50–65	66–80	81–88
Areas 1, 3	4	0–16	0–4	5–7	8–10	11–13	14–15	16
Areas 2, 4, 5	3	0–12	0–3	4–5	6–7	8–9	10–11	12
Area 6	5	0–20	0–6	7–8	9–12	13–16	17–19	20

and adaptation for the Greek context are both valid and reliable in helping teachers self-assess their digital competence level.

2.3 Data Analysis Procedure

The Greek version of the DigCompEdu self-assessment questionnaire was disseminated online through the EuSurvey platform. This platform supports the provision of direct, personalized feedback to each participant upon completing the questionnaire, helping teachers reflect on their personal strengths and weaknesses regarding their perceived level of digital competence across the six areas that make up the DigCompEdu framework. Participation in the study was anonymous and voluntary, and the data collected from the respondents was kept confidential and used solely for research purposes.

The reliability and convergent validity of the questionnaire were assessed using the following coefficients: Cronbach's alpha, McDonald's omega, Composite reliability (CR), and Average variance extracted (AVE). To determine the construct validity of the questionnaire, an Exploratory Factor Analysis (EFA) was conducted. The principal components method with varimax rotation and Kaiser normalization was employed for factor selection. After determining the number of factors, a Confirmatory Factor Analysis (CFA) was performed to verify whether the data fit the hypothesized measurement model derived from the Exploratory Factor Analysis using Structural Equation Modeling (SEM). The maximum likelihood method was applied to test the theoretical model. The overall goodness-of-fit of the factor model was evaluated using the following indices: CMIN, CFI, GFI, NFI, PNFI, RMR, and RMSEA [17]. Following this, a descriptive analysis of central tendency (mean) and dispersion (standard deviation) was conducted to determine the level of teachers' digital competence both overall and for each dimension that constitutes the DigCompEdu framework.

The IBM SPSS (v.25) statistical software and the IBM SPSS Amos (v.23) structural equation modeling (SEM) software were used for data analysis.

3 Results

3.1 Exploratory Factor Analysis (EFA)

Exploratory Factor Analysis (EFA) is a multivariate statistical technique widely used in the development and validation of measurement tools to identify underlying relationships among measured variables and to determine the factor structure of a scale [18]. In this study, EFA was performed to determine the factorial validity of the

Greek version of the DigCompEdu Check-In self-assessment questionnaire. Principal component analysis with varimax rotation was used as the extraction method. Prior to this, its applicability for factor analysis was confirmed based on standard indicators. Bartlett's Test of Sphericity (BTS) was conducted to examine the interdependence among the items. Furthermore, the Kaiser-Meyer-Olkin (KMO) sampling statistic test was calculated to measure the sampling adequacy [19].

Table 2 presents the results of factor analysis that confirm the questionnaire structure's relevance. All factors are valid and reliable structures, and all variables are suitable items for measuring them. More specifically, Bartlett's Test of Sphericity (BTS) yielded a significant result ($p < 0.001$) for all factors, indicating adequate intercorrelations among the variables. The Kaiser-Meyer-Olkin (KMO) index is greater than 0.600, which exceeds the recommended value of 0.50 [20], indicating the suitability of the dataset for conducting factor analysis. All factor loadings are greater than the cut-off point 0.550 and the Total Variance Explained (TVE) index is greater than 50.000 for all factors, indicating a robust factor structure [17]. The six factors identified as part of this EFA corresponded with the six theoretical factors proposed in this research. Therefore, the factor structure proposed by EFA was tested using CFA.

3.2 Confirmatory Factor Analysis (CFA)

The proposed fix-factor structure that emerged as a result of the Exploratory Factor Analysis (EFA) was tested using Confirmatory Factor Analysis (CFA). This method is used to confirm a predefined factor structure based on theoretical assumptions or prior knowledge, such as the EFA results. Contrary to EFA, CFA enables researchers to define the number of factors and the existing relationships between observed and latent variables. It is a deductive approach that helps ensure that the measurement model aligns with the theoretical construct [17]. Furthermore, several authors underscore the suitability of this confirmatory analysis for verifying the validity of an instrument adapted to another language or applied to a distinct population, as it cannot be guaranteed that the measurement model is robust and applicable in the same way and retain the same conceptual meaning across diverse groups [21].

The IBM SPSS AMOS 24.0 software package was used to test the twenty-two (22) measurement items with six (6) factor structures. The factorial validity of the instrument was assessed through a Confirmatory Factor Analysis with maximum likelihood method. The aim is to confirm the variables and the structure of the scale, showing the existing relationships between the different dimensions. Figure 2 presents the CFA results, showing the proposed structural diagram with item-dimension and dimension-dimension correlation indices.

As illustrated in the model, the six latent variables (A-F) are properly correlated with their items. Factor loadings of the scale items ranged from 0.51 to 0.87, with 15 of 22 items meeting or exceeding the ideal threshold of 0.70, and the remaining 7 items fell within the acceptable range of 0.50–0.70 [22]. None of the items exhibited

Table 2 Exploratory factor analysis results

Items	F_A	F_B	F_C	F_D	F_E	F_F	KMO	BTS	TVE
A1	0.744						0.742	$p < 0.001$	51.851
A2	0.727								
A3	0.723								
A4	0.686								
B1		0.828					0.612	$p < 0.001$	58.040
B2		0.773							
B3		0.676							
C1			0.845				0.796	$p < 0.001$	64.397
C2			0.789						
C3			0.729						
C4			0.842						
D1				0.878			0.620	$p < 0.001$	68.760
D2				0.693					
D3				0.901					
E1					0.829		0.678	$p < 0.001$	65.633
E2					0.772				
E3					0.828				
F1						0.782	0.837	$p < 0.001$	63.117
F2						0.799			
F3						0.794			
F4						0.779			
F5						0.818			

Extraction method: principal component analysis
F_A professional engagement, *F_B* digital resources, *F_C* teaching and learning, *F_D* assessment, *F_E* empowering learners, *F_F* facilitating learners' digital competence

a factor loading below 0.50, indicating that all items contribute adequately to their respective factors.

The goodness-of-fit indices were calculated to evaluate the quality of the model. Table 3 shows the obtained and the reference values for the fit of the model according to [17] along with the recommended thresholds for accepted scores: Chi-Squared (CMIN), comparative fit index (CFI), good-ness-of-fit index (GFI), normed fit index (NFI), parsimony normed fit index (PNFI), root mean square residual (RMR), root mean square error of approximation (RMSEA).

Considering the indices obtained, it has been confirmed that the proposed measurement model is adequate and fits perfectly with the empirical data, as all the indices meet or exceed the recommended thresholds, demonstrating that the proposed model is adequate and fits perfectly with the empirical data. These findings supported and validated the six-factor structure proposed in the EFA.

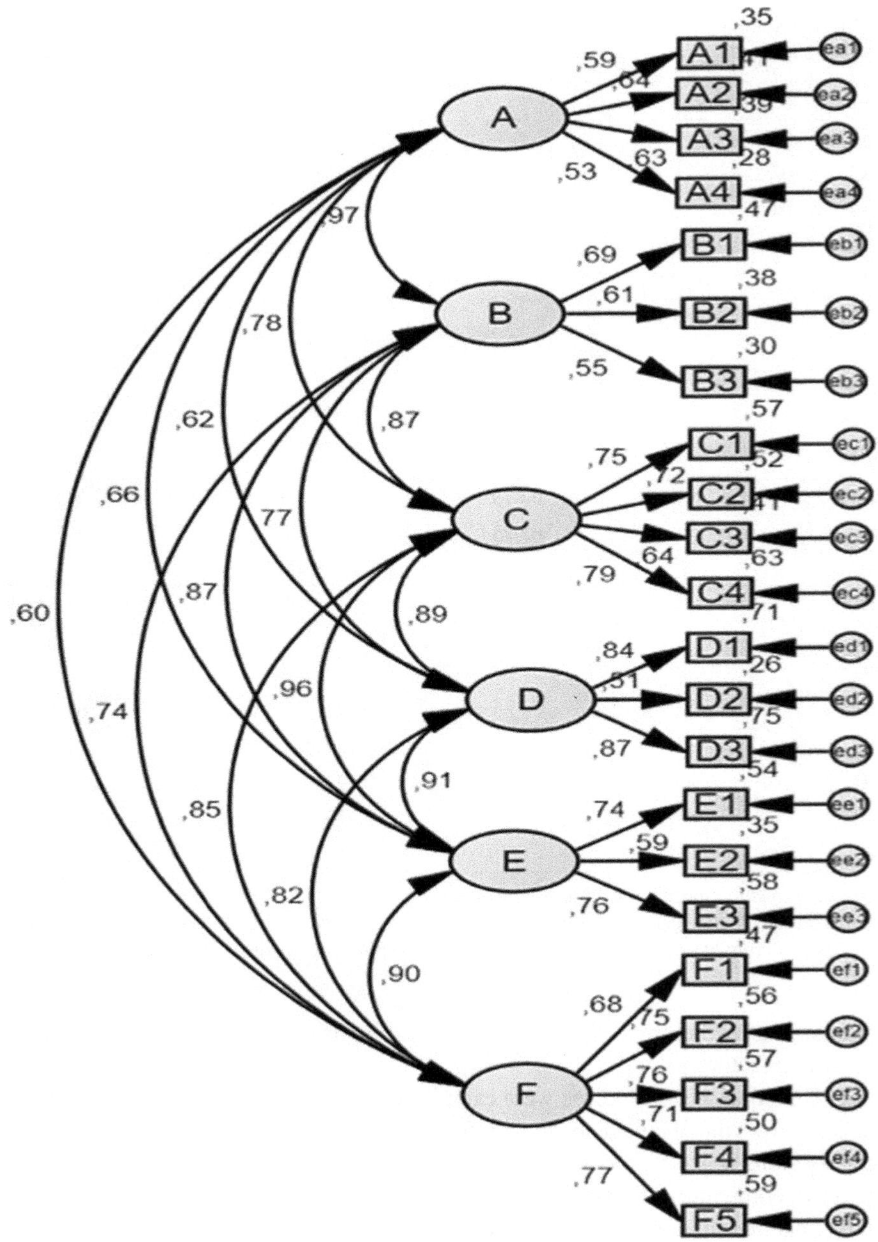

Fig. 2 CFA results—structure diagram of the Greek version of the "DigCompEdu Check-In" self-assessment questionnaire

Table 3 Goodness-of-fit indices of the model

Index	Result	Accepted scores	Fit
CMIN/DF	1.481	< 5	Excellent
CFI	0.956	> 0.9	Excellent
GFI	0.888	> 0.8	Good
NFI	0.878	> 0.8	Good
PNFI	0.738	> 0.7	Good
RMR	0.061	< 0.07	Good
RMSEA	0.048	< 0.07	Excellent

3.3 Reliability and Convergent Validity Analysis

The reliability of the Greek version of the DigCompEdu Check-In self-assessment questionnaire was assessed using multiple reliability indices to ensure the psychometric soundness of the scale. Cronbach's alpha (α) coefficient and McDonald's omega (ω) coefficient were calculated globally and for each dimension of the questionnaire to measure its internal consistency. Table 4 presents the obtained levels of reliability according to the psychometric measures employed.

According to [23], Alpha (α) and Omega (ω) values greater than 0.7 are considered sufficient to ensure the reliability of the instrument, both globally and for each of their dimensions. The entire questionnaire with 22 items has a high internal consistency for both indices ($a = 0.933$) and ($\omega = 0.933$). When examining the internal consistency by dimension, values for both indices range from 0.622 to 0.853. This range is considered acceptable to good, with the exception of Dimension A (Professional engagement) and Dimension B (Digital Resources), which fall slightly below the threshold of 0.7 for both indices. These results can still be considered acceptable within certain research contexts. Previous work has suggested that values as low as 0.600 are acceptable when measuring subjective constructs, such as estimations

Table 4 Reliability statistics

Dimension	Cronbach's Alpha	McDonald's Omega	Number of elements
D_A: Professional engagement	0.685	0.684	4
D_B: Digital resources	0.622	0.630	3
D_C: Teaching and learning	0.811	0.818	4
D_D: Assessment and feedback	0.756	0.772	3
D_E: Empowering learners	0.727	0.728	3
D_F: Facilitating learners' digital competence	0.852	0.851	5
Total	0.933	0.933	22

Table 5 Convergent validity of the model

Dimension	CR	Adjust	AVE	Adjust
D_A: Professional engagement	0.812	CR > 0.7	0.519	AVE > 0.5
D_B: Digital resources	0.805		0.580	
D_C: Teaching and learning	0.878		0.644	
D_D: Assessment and feedback	0.867		0.688	
D_E: Empowering learners	0.851		0.657	
D_F: Facilitating learners' digital competence	0.895		0.631	

of opinion or critique. Similarly, lower reliability coefficients are reasonable when questionnaires are applied in varied contexts or populations, as noted in [24]. Considering the exploratory nature of this study and the adaptation of the questionnaire in the Greek context, the slightly lower reliability coefficient for these dimensions is not unexpected and does not undermine the overall validity of the instrument.

Convergent validity is a key component of construct validity. It refers to the extent to which items or indicators designed to measure a particular construct (e.g., a concept like "digital competence") are strongly related to the theoretical construct they are intended to measure [25]. To determine the convergent validity of the questionnaire, two statistical indicators were employed: Composite Reliability (CR) and Average Variance Extracted (AVE). As noted in [17], convergent validity is observed when CR is higher than AVE and AVE is higher than 0.5. Table 5 shows the results, as well as the reference values obtained for model adjustment. As observed, CR scores ranged from 0.805 to 0.895, exceeding the threshold of 0.7, and all AVE values surpassed 0.5, demonstrating the indicators' valid representation of the construct.

These findings suggest that the Greek version of the DigCompEdu Check-In self-assessment questionnaire demonstrates adequate convergent validity, as all the aforementioned thresholds are met in this study, and the items within each construct are highly correlated, reflecting the same underlying concept.

3.4 Description of the Level of Digital Teaching Competence of Greek Teachers

Participants were asked to self-evaluate their level of digital competence be-fore and after completing the DigCompEdu Check-In questionnaire. This self-evaluation was based on the progression model proposed by the DigCompEdu framework, which aligns with the six proficiency levels of the Common European Framework of Reference for Languages (CEFR), ranging from A1 to C2. Complementary to the pre- and post-self-ratings, the questionnaire also provides a total score for each participant.

Table 6 presents the mean values and standard deviations of participants' self-perceived level of digital competence before and after completing the questionnaire, as well as their level of digital competence based on the score obtained from the questionnaire. To ensure correct interpretation of these data, it is important to note that the response interval ranges from 1 to 6, representing six response options, with 1 indicating the lowest level (A1) and 6 the highest level (C2). As can be observed, the average self-perceived DTC of the participants was 2.82 (SD = 1.117) before completing the questionnaire, while after completing the questionnaire, the average value in-creased to 2.87 (SD = 1.060), indicating that participants positioned them-selves slightly higher than previously. Therefore, their perception of their DTC improved after completing the questionnaire. Notably, the level of DTC based on the score obtained from the questionnaire was higher than the self-reported pre- and post-ratings of the participants (mean = 3.30, SD = 1.081).

The descriptive analysis is completed with the information presented in Table 7, which displays the mean values and standard deviations reported by the participating teachers for the six (6) competence areas and the twenty-two (22) specific competences that constitute the DigCompEdu Check-In questionnaire. To ensure that these data are interpreted accurately, it is important to note that the response interval ranges from 1 to 5, representing a total of five response options, with 1 indicating the lowest and 5 the highest possible score. Overall, the results suggest that teachers demonstrate average levels of competence across all dimensions. Ordered from lowest to highest, the results by areas of competence are as follows:

Area 3 (D_C): Teaching and learning (3.03).
Area 6 (D_F): Facilitating learners' digital competence (3.05).
Area 4 (D_D): Assessment and feedback (3.09).
Area 1 (D_A): Professional engagement (3.13).
Area 2 (D_B): Digital resources (3.20).
Area 5 (D_E): Empowering learners (3.33).

Notably, Area 5 (Empowering Learners) has the highest average score, suggesting that teachers feel most confident in supporting inclusion, differentiation, and actively engaging learners in the teaching process. Conversely, the least developed areas include Area 3 (Teaching and Learning), Area 6 (Facilitating Learners' Digital Competence), and Area 4 (Assessment and Feedback), indicating that Greek teachers will need to invest more and may benefit from targeted training in these specific areas.

Table 6 Overall Digital Teaching Competence of Greek teachers based on the DigCompEdu framework

	Mean	SD
Self-perceived level of DTC level (before)	2.82	1.117
Self-perceived level of DTC (after)	2.87	1.060
DTC based on the questionnaire's total score	3.30	1.081

Table 7 Descriptive statistics of Greek teachers' digital competence according to the six competence areas and the twenty-two specific competences of the DigCompEdu framework

Competence area	Specific competence	Mean per item	SD per item	Mean per area	SD per area
D_A: Professional engagement	A1. Organizational communication	3.09	1.010	3.13	1.100
	A2. Professional collaboration	2.79	1.026		
	A3. Reflective practice	3.17	1.093		
	A4. Digital CPD	3.47	1.255		
D_B: Digital resources	B1. Selecting digital resources	3.12	0.940	3.20	1.090
	B2. Creating and modifying digital resources	3.25	1.133		
	B3. Managing, protecting and sharing digital resources	3.23	1.169		
D_C: Teaching and learning	C1. Teaching	2.85	1.222	3.03	1.210
	C2. Guidance	2.96	1.393		
	C3. Collaborative learning	3.21	1.050		
	C4. Self-regulated learning	3.09	1.156		
D_D: Assessment and feedback	D1. Assessment strategies	3.03	0.884	3.09	0.960
	D2. Analyzing evidence	3.23	1.060		
	D3. Feedback and planning	3.00	0.930		
D_E: Empowering learners	E1. Accessibility and inclusion	3.35	1.124	3.33	1.140
	E2. Differentiation and personalization	3.38	1.291		

(continued)

Table 7 (continued)

Competence area	Specific competence	Mean per item	SD per item	Mean per area	SD per area
	E3. Actively engaging learners	3.26	0.973		
D_F: Facilitating learners' digital competence	F1. Information and media literacy	3.23	1.186	3.05	1.100
	F2. Digital communication	2.76	1.158		
	F3. Digital content creation	2.93	1.005		
	F4. Responsible use	3.20	1.056		
	F5. Digital problem solving	3.15	1.027		

4 Discussion and Conclusions

As a starting point, the present study aimed to assess the reliability and validity of the Greek version of the DigCompEdu Check-In self-assessment questionnaire and to determine the level of digital competence of the participants, both overall and by competence areas, according to the DigCompEdu framework. The study also sought to identify teachers' professional development needs in the least developed areas of the DigCompEdu framework to plan targeted professional development pathways for continuous improvement.

The DigCompEdu Check-In self-assessment questionnaire has been validated in various languages and contexts. Several previous empirical studies have thoroughly validated this instrument in Germany [8], in Morocco [26], in Spain [27], in Portugal [28], and in Hungary [29]. As shown by the results obtained from this study, the Greek version of this self-assessment instrument demonstrates high reliability indices, both globally and across all its dimensions. These findings allow for comparisons with results obtained in similar studies. It is worth mentioning that, in our study, the internal consistency reliability of Area 2 (Digital Resources) falls slightly below the threshold of 0.7, as suggested by the rule of thumb provided by [17]. A similar potential reliability issue regarding Area 2 (Digital Resources) has also been reported in a German sample [8] and in a case study with Moroccan teachers [26]. Nevertheless, both studies confirmed the validity and reliability of the self-assessment instrument. As the results obtained in our study regarding the reliability of this area fall slightly below the threshold of 0.7, we do not attempt to remove items from this area, and we acknowledge that lower reliability coefficients can be considered valid and reasonable when questionnaires are applied in varied contexts or populations.

Even though many researchers have found that this self-assessment tool exhibits high levels of internal consistency, digital competence is a multidimensional construct, and further analysis is needed to explore and validate the factorial structure of the DigCompEdu Check-In self-assessment questionnaire [8]. The study in [24] conducted exploratory and Confirmatory Factor Analysis using structural equation modeling, confirming the original six-factor structure of the self-assessment questionnaire. Conversely, the study in [30] applied the same techniques, resulting in a one-factor structure. In our study, the results of the Confirmatory Factor Analysis validate the theoretical six-factor structure model as outlined in the DigCompEdu framework. Based on the fit indices obtained, the model demonstrates an excellent fit with the empirical data and establishes adequate convergent validity.

Regarding the overall level of teachers' digital competence, the study sample demonstrates a medium level of teaching digital competence, with most teachers falling into the B1 (Integrator) and B2 (Expert) proficiency levels. These results are consistent with previous studies [8, 26–28]. It is worth mentioning that the results reveal a mismatch between the self-perceived level of teachers' digital competence and their actual digital teaching competence based on the scores obtained from the questionnaire. More specifically, the participants in our study tend to underestimate their digital competence compared to the actual scores obtained from the questionnaire. These findings align with the results reported by [8], but contradict those of [29], where participants were found to overestimate their level of digital competence compared to the actual score obtained from the questionnaire. Examining teachers' digital teaching competence by areas of competence, it is noteworthy that the lowest levels of competence exhibited by the Greek teachers participating in this study were in Area 3 (Teaching and Learning), Area 6 (Facilitating Learners' Digital Competence), and Area 4 (Assessment and Feedback). These findings under-score the need for teachers to enhance their digital competence levels in these areas, particularly in the pedagogical use of digital technologies.

The limitations of this study should be acknowledged. First, there is a selection bias, as participation in the study was voluntary, and the sampling method chosen was non-probabilistic. Consequently, it is possible that the sample is not representative, as it is likely that those who chose to participate may differ significantly from those who did not. Additionally, the number of participating teachers ($N = 208$) is relatively small, particularly in comparison with the total number of teachers in Greek primary and secondary education. However, since the aim of this study was exploratory, focusing on validating of the Greek version of the DigCompEdu self-assessment questionnaire, the findings cannot be generalized to the entire population of Greek teachers. Nevertheless, the sample obtained allows us to draw some general conclusions regarding their training needs in digital competence and draw insights regarding the reliability of the Greek version of this self-assessment questionnaire and its underlying factorial structure. Finally, in addition to selection bias, the fact that the study referred to self-reported data may introduce further limitations as it reflects teachers' self-perception rather than the actual data regarding their level of digital competence. To overcome this limitation, it should be useful to supplement teachers' self-reports with knowledge-based tests designed to guarantee an objective assessment of the actual digital teaching competence.

The findings of this study can inform the design of targeted professional development programs in Greece, focusing on the areas where teachers demonstrated lower competence. Researchers, educational policymakers and school administrators can use the validated tool to assess digital training needs and provide personalized training pathways aligned with the DigCompEdu framework. As future research lines or projects it would be worth considering the creation of a TPD-MOOC focused on teaching and learning in online and blended inclusive learning environments in the digital age. This training proposal could offer free professional development opportunities to teachers at all levels of compulsory education, with the aim to enhance their digital competence in the least developed areas of the DigCompEdu that require further improvement.

Acknowledgments The research work was supported by the Hellenic Foundation for Research and Innovation (HFRI) under the third Call for HFRI PhD Fellowships (Fellowship Number: 06513).

References

1. Kusnadi, K., Purwantiningsih, A., Y.: Sudarso: digital society based on public civility: an alternative to strengthen national character. J. Educ. Hum. Soc. Sci. **6**(1), 328–342 (2023). https://doi.org/10.34007/jehss.v6i1.1872
2. European Commission: Council recommendation on key competences for lifelong learning. Off. J. Eur. Union. **189**, 1–13 (2018)
3. Ferrari, A.: Digital Competence in Practice: An Analysis of Frameworks. Publications Office of the European Union, Luxembourg (2012). https://data.europa.eu/doi/10.2791/82116
4. Wild, S., Schulze Heuling, L.: How do the digital competences of students in vocational schools differ from those of students in cooperative higher education institutions in Germany? Empirical research in vocational. Educ. Train. **12**(1), 5 (2020). https://doi.org/10.1186/s40461-020-00091-y
5. Ilomäki, L., Paavola, S., Lakkala, M., Kantosalo, A.: Digital competence—an emergent boundary concept for policy and educational research. Educ. Inf. Technol. **21**, 655–679 (2016). https://doi.org/10.1007/s10639-014-9346-4
6. Cattaneo, A.A., Antonietti, C., Rauseo, M.: How digitalized are vocational teachers? Assessing digital competence in vocational education and looking at its underlying factors. Comput. Educ. **176**, 104358 (2022). https://doi.org/10.1016/j.compedu.2021.104358
7. Admiraal, W., van Vugt, F., Kranenburg, F., Koster, B., Smit, B., Weijers, S., Lockhorst, D.: Preparing pre-service teachers to integrate technology into K–12 instruction: evaluation of a technology-infused approach. Technol. Pedagog. Educ. **26**(1), 105–120 (2017). https://doi.org/10.1080/1475939X.2016.1163283
8. Ghomi, M., Redecker, C.: Digital Competence of Educators (DigCompEdu): Development and evaluation of a self-assessment instrument for teachers' digital competence. In: Proceedings of the 11th International Conference on Computer Supported Education—Volume 1: CSEDU, pp. 541–548. Science and Technology Publications (2019). https://doi.org/10.5220/0007679005410548
9. Engen, B.K.: Understanding social and cultural aspects of teachers' digital competencies. Comunicar Media Educ. Res. J. **27**(61), 9–18 (2019). https://doi.org/10.3916/C61-2019-01

10. Pettersson, F.: On the issues of digital competence in educational contexts—a review of literature. Educ. Inf. Technol. **23**(3), 1005–1021 (2018). https://doi.org/10.1007/s10639-017-9649-3
11. McGarr, O., McDonagh, A.: Digital competence in teacher education. Output 1 of the Erasmus+ funded Developing Student Teachers' Digital Competence (DICTE) project. https://hdl.handle.net/10344/7700 (2019)
12. Cabero-Almenara, J., Romero-Tena, R., Palacios-Rodríguez, A.: Evaluation of teacher digital competence frameworks through expert judgement: the use of the expert competence coefficient. J. New Approach. Educ. Res. **9**(2), 275–293 (2020). https://doi.org/10.7821/naer.2020.7.578
13. Redecker, C., Punie, Y.: European Framework for the Digital Competence of Educators: DigCompEdu. Publications Office of the European Union, Luxembourg (2017)
14. Caena, F., Redecker, C.: Aligning teacher competence frameworks to 21st century challenges: the case for the European digital competence framework for educators (Digcompedu). Eur. J. Educ. **54**(3), 356–369 (2019). https://doi.org/10.1111/ejed.12345
15. Vergidis, D.: Educational needs (educational needs assessment). Adult Educ. Crit. Issues. **1**(1), 78–80 (2021)
16. Creswell, J.W.: Research Design: Qualitative, Quantitative, and Mixed Methods Approaches, 4th edn. Sage Publications, Thousand Oaks (2014)
17. Hair, J.F., Black, W.C., Babin, B.J., Anderson, R.E.: Multivariate Data Analysis: a Global Perspective. Pearson Education, Upper Saddle River (2010)
18. Watkins, M.W.: Exploratory factor analysis: a guide to best practice. J. Black Psychol. **44**(3), 219–246 (2018). https://doi.org/10.1177/0095798418771807
19. Field, A.: Discovering Statistics Using SPSS. Sage Publications Ltd., London (2005)
20. Clark, L.A., Watson, D.: Constructing validity: basic issues in objective scale development. In: Kazdin, A.E. (ed.) Methodological Issues and Strategies in Clinical Research, 4th edn, pp. 187–203. American Psychological Association, Washington (2016). https://doi.org/10.1037/14805-012
21. Beauford, J.E., Nagashima, Y., Wu, M.H.: Using translated instruments in research. J. College Teach. Learn. **6**(5) (2009). https://doi.org/10.19030/tlc.v6i5.1144
22. Cheung, G.W., Cooper-Thomas, H.D., Lau, R.S., Wang, L.C.: Reporting reliability, convergent and discriminant validity with structural equation modeling: a review and best-practice recommendations. Asia Pac. J. Manag. **41**(2), 745–783 (2024). https://doi.org/10.1007/s10490-023-09871-y
23. O'Dwyer, L.M., Bernauer, J.A.: Quantitative Research for the Qualitative Researcher. SAGE Publications, Thousand Oaks (2013)
24. Llorente-Cejudo, C., Barragán-Sánchez, R., Puig-Gutiérrez, M., Romero-Tena, R.: Social inclusion as a perspective for the validation of the "DigCompEdu check-in" questionnaire for teaching digital competence. Educ. Inf. Technol. **28**(8), 9437–9458 (2023). https://doi.org/10.1007/s10639-022-11273-4
25. Chin, C.L., Yao, G.: Convergent validity. In: Maggino, F. (ed.) Encyclopedia of Quality of Life and Well-being Research. Springer, Cham (2023). https://doi.org/10.1007/978-3-031-17299-1_573
26. Benali, M., Kaddouri, M., Azzimani, T.: Digital competence of Moroccan teachers of English. Int. J. Educ. Dev. Using Inf. Commun. Technol. **14**(2), 99–120 (2018)
27. Cabero-Almenara, J., Gutiérrez-Castillo, J.J., Palacios-Rodríguez, A., Barroso-Osuna, J.: Development of the teacher digital competence validation of DigCompEdu check-in questionnaire in the university context of Andalusia (Spain). Sustainability. **12**(15), 6094 (2020). https://doi.org/10.3390/su12156094
28. Dias-Trindade, S., Moreira, J.A., Nunes, C.S.: Self-evaluation scale of teachers' digital competences. Construction and validation procedures. Texto Livre. **12**(2), 152–171 (2019). https://doi.org/10.17851/1983-3652.12.2.152-171
29. Horváth, L., Pintér, T.M., Misley, H., Dringó-Horváth, I.: Validity evidence regarding the use of DigCompEdu as a self-reflection tool: the case of Hungarian teacher educators. Educ. Inf. Technol. (2024). https://doi.org/10.1007/s10639-024-12914-6

30. Martín Párraga, L., Llorente Cejudo, C., Barroso Osuna, J.: Validation of the DigCompEdu check-in questionnaire through structural equations: a study at a university in Peru. Educ. Sci. **12**(8), 574 (2022). https://doi.org/10.3390/educsci12080574

Teachers' Beliefs and Pedagogical Thinking About Flipped Learning

Anna Dimitrakopoulou and Athanassios Jimoyiannis

Abstract This paper presents the results of a study exploring teachers' beliefs and pedagogical perceptions of flipped learning as well as their readiness to adopt and implement flipped learning interventions in their instructional practices. Quantitative research data were collected through an online questionnaire completed by 283 primary and secondary school teachers serving in Greek public schools. Data analysis showed that most teachers were positive about using the flipped learning approach in their classrooms. They also recognize the key pedagogical features of the flipped learning model, the consequent changes in the process of learning, and the benefits for the students. Conversely, the findings indicated that teachers' practical experience with the flipped learning model and teachers' training in using ICT in their instruction positively influence their perceptions about flipped learning.

Keywords Flipped learning · Flipped classroom · ICT in education · K-12 education · Teachers' perceptions

1 Introduction

In recent years the interest in flipped learning (FL) has grown significantly among K12 teachers and higher education instructors as they strive to implement innovative teaching strategies that enhance students' engagement and improve learning outcomes to align with twenty-first-century society's needs. The flipped learning model is built on the concept of shifting the teaching phase outside the classroom—typically at home—before in-class sessions, thereby redistributing students' cognitive load. The flipped learning model aligns with the four pillars of F-L-I-P: Flexible Environment, Learning Culture, Intentional Content, and Professional Educator.

A. Dimitrakopoulou · A. Jimoyiannis (✉)
Department of Social and Educational Policy, University of Peloponnese, Korinthos, Greece
e-mail: ajimoyia@uop.gr

A. Dimitrakopoulou
e-mail: dimi_anna@hotmail.com

I. Kazanidis and A. Tsinakos (eds.), *Research on E-Learning and ICT in Education*,
https://doi.org/10.1007/978-3-031-99961-1_6

This framework emphasizes creating adaptable learning spaces, fostering a student-centered culture, thoughtfully selecting content, and maintaining an active, guiding role for educators [1].

The fundamental difference between the traditional classroom and the flipped classroom lies in the inversion of the teaching-learning sequence. In the FL model, students engage with instructional content prior to class, utilizing various materials such as textbooks, video lessons, and digital resources, often accessible through a Learning Management System (LMS) (e.g., eClass). This approach liberates in-class time, allowing students to actively participate in creative, constructive, and collaborative learning experiences under the teacher's guidance [2].

Therefore, flipped learning moves direct instruction from the group learning space (in-class space and time) into an individual learning space (before class) where students are expected to explore and familiarize themselves with the basic knowledge. On the other side, the classroom is transformed into an active learning space created through students' interaction and collaboration, exploration, application of new knowledge, and problem-solving. By restructuring the traditional instructional sequence, the FL model promotes active learning and enhances student engagement, leading to improved educational outcomes. To successfully implement this shift, teachers require adequate support and training. Professional development programs can equip educators with the necessary skills to produce engaging instructional videos, design interactive and meaningful learning activities, and foster a student-centered approach to education [3, 4].

Flipped learning has been widely used in higher education [5, 6]. Rooted in its open, creative, and student-centered features, flipped learning has been also proposed as an effective educational practice for school education. Thus, FL interventions have been implemented across various subjects in primary and secondary education, such as mathematics [7–9], science [10, 11], English language [12–14], humanities, and social sciences [15].

The literature review indicated that the majority of current research focuses on students' adoption, learning outcomes, and satisfaction with the FL model [16–19]. Overall, students tend to have a positive view of the FL approach and acknowledge several benefits regarding their learning outcomes [13, 19–21]. In contrast, research on teachers' views and their readiness to adopt the FL model, as well as to design and incorporate FL interventions into their teaching practices, is still relatively scarce [18, 22, 23]. The findings of two recent systematic reviews on flipped learning in K-12 education [16, 24] showed that research has primarily concentrated on student perceptions of flipped learning, their engagement, and performance in STEM subjects. Moreover, these studies have mostly been conducted in high schools in North America and Asia.

Although the educational interest about FL has significantly increased in recent years, teachers' perceptions and readiness to adopt the FL model, as well as their abilities to design and support flipped learning in their class, remain an open research issue internationally [5, 23, 24]. Addressing the need for further research, this paper presents preliminary findings from a quantitative study examining Greek primary and secondary teachers' perceptions regarding the pedagogical aspects of FL, their abilities to design and implement flipped learning interventions in educational settings, as well as the challenges and barriers they perceive in adopting the FL model.

2 Rationale of the Study

Empirical research worldwide has consistently highlighted the benefits of the flipped learning (FL) approach in creating a more flexible and active learning environment [25]. Studies indicate that FL positively influences student motivation, autonomous learning, peer interaction, collaboration, and overall satisfaction [19, 26]. Additionally, research supports that this instructional model encourages participation in classroom discussions, promotes collaborative learning, and enhances students' learning habits and attitudes [9, 27, 28]. FL has also been found to improve communication and problem-solving skills, ultimately leading to better learning outcomes and academic performance compared to traditional lecture-based teaching methods [7, 29–32]. Beyond improving student outcomes, FL shifts the teacher's role from a lecturer to a facilitator and mediator of the learning process, fostering a more student-centered educational experience [33]. Moreover, it contributes to the development of students' technology literacy, metacognitive thinking skills [34], enabling a deeper learning experience [35].

In general, students appear positive about the FL approach, and they recognize many benefits in terms of their learning achievements [21, 36]. On the other side, teachers encounter many barriers in organizing and implementing flipped classroom instruction, since it is opposed to traditional teaching, it is time-consuming and requires increased workload from their part [37–39]. Additionally, they often lack appropriate learning resources and open educational materials for their FL interventions, such as video lessons and digital content required for students' preparation during the pre-class phase [38]. Other frequently mentioned factors are student-related, including students' effective pre-class preparation and use of the educational material provided [37, 39–41] and students' limited access to online technologies at home [20, 33, 42].

Cheng et al. [36] reported that a serious difficulty for the teachers is that many students are initially unreceptive of the FL model, and they lack self-regulation skills that are necessary for effective FL interventions. As Wang [41] advocated, successful FL interventions require motivated and confident teachers who need to put high initial effort and have access to appropriate resources to develop effective FL designs in practice. Consequently, many teachers may be reluctant to adopt the FL approach and resist change due to their beliefs, attitudes, and lack of confidence in digital skills.

Some studies have found that educators struggle to implement flipped learning interventions due to a lack of the necessary knowledge and skills required to effectively integrate and support the FL model in their educational practices [5]. The successful implementation of the FL model requires teachers to develop sufficient technological and pedagogical knowledge to facilitate learning in both pre-class and face-to-face settings [43]. According to Chou et al. [22], implementing the FL approach requires a supportive school environment that provides the necessary infrastructure as well as professional development opportunities for the teachers to enhance their ability to prepare and organize FL interventions.

A significant challenge for teachers is to understand the pedagogical meaning of "flipping" in its overall depth. FL cannot be achieved without the integration of digital and online technologies. Thus, many teachers, under growing pressure to incorporate digital tools into their instruction, may view FL merely as a trend or a teaching method enhanced by technology. It's important to note that this perspective lacks true pedagogical meaning or reflection. Flipped learning also involves a shift from teacher-centered to student-centered learning. Thus, FL is closely linked to a variety of student-centered pedagogical approaches, such as active learning, inquiry-based learning, problem-based learning, case-based learning, and collaborative learning [24, 40, 44, 45].

Moreover, teachers may view flipped learning as a type of blended learning, as it integrates traditional classroom settings with online environments and interactive digital activities that promote active student engagement. A deeper connection between flipped and blended learning is established through the integration of multiple learning theories, including behaviorism, cognitivism, constructivism, social learning theory, and connectivism [16, 46].

The findings of a recent qualitative research study in Greek schools indicated that teachers recognize several fundamental pedagogical features as essential to the flipped learning model [40]: (a) active learning; (b) better use of in-class time; (c) students' interaction and collaboration; (d) development of high-level learning skills; (e) deepening and applying new knowledge; (f) creative learning; (g) development of twenty-first-century skills; (h) personalized/differentiated learning. In the same vein, a recent study in Finnish secondary schools found that mathematics teachers viewed FL as a suitable pedagogical approach for all students, highlighting three key pedagogical features: individualized learning, self-regulated learning, and active learning to enhance student engagement [47].

In conclusion, existing research shows that most teachers face many difficulties, and they are cautious to implement flipped learning interventions in practice. The main factors that introduce challenges to the teachers and affect the adoption of FL are: (a) lack of technological infrastructure, (b) increased workload, (c) student resistance and negative attitudes/inadequate preparation, (d) lack of support from the broader educational environment (leadership, parents), (e) lack of skills in instructional design and proper professional development for teachers.

2.1 Research Context and Research Questions

Our interest in exploring teachers' beliefs about FL was based on the growing adoption of digital tools and online practices in school education, which has been observed in Greek schools over the recent years. The rapid shift to emergency remote online teaching, when schools closed due to the pandemic of COVID-19, forced many teachers to experiment with new pedagogical ideas, such as the FL approach [8, 48]. Additionally, the findings of a previous study involving Greek teachers revealed that the flipped learning approach was the most widely adopted instructional method during the pandemic [49].

The hypothesis of this study was that many primary and secondary education teachers would be familiar with the FL model and willing to participate in a study regarding their views about the pedagogical features of the model. Responding to this objective, the following research questions were investigated:

- What are the teachers' beliefs regarding the pedagogical features of the FL model and the changes it brings into the learning process?
- What are the teachers' beliefs about their abilities to design and implement effective FL interventions? What are their concerns and difficulties?
- Are there statistically significant differences related to the demographic characteristics of teachers (e.g., gender, level of education, experience, training)?

3 Research Method

The present study adopted a mixed research method, collecting both quantitative and qualitative data regarding teachers' views and beliefs about the pedagogical features of FL, as well as the challenges and difficulties they face in designing and supporting FL practices. However, in this paper we present the results of the quantitative data analysis.

The data collection period lasted for 3 weeks, and a snowball sampling method was chosen as a form of convenience sampling. We used an online questionnaire via Google Forms to collect both quantitative and qualitative data regarding teachers' beliefs about FL.

3.1 The Sample

A total of 283 fully completed questionnaires were included in the quantitative analysis. The sample comprised primary and secondary school teachers serving in public schools across various regions of the country. Among the participants, 71 (25.1%) were men, and 212 (74.9%) were women; 56.9% taught in primary schools, while 43.1% taught in high schools. The majority (51%) had over 20 years of teaching experience.

Approximately six out of ten teachers (58.3%) reported having used the flipped learning (FL) approach in their instruction at least once. Additionally, 48% of the teachers attended a specific seminar on flipped learning. Furthermore, 113 teachers participated in a 36-hour professional development program focused on educational innovations and twenty-first-century skills. This program, offered by the Institute of Educational Policy under the aegis of the Greek Ministry of Education, included flipped learning as a distinct thematic unit.

Most teachers in the sample reported possessing strong digital skills and competence in using ICT in education, a result attributed to their participation in the three-tiered National In-service Teacher Training Program (NITTP). This program was designed to enhance teachers' knowledge and skills for effectively integrating ICT into classroom practice. Among the participants, 39.9% reported attaining only the basic A-level qualification, 33% achieved the B1 level, and 35% reached also the B2 level.

3.2 The Scale

The development of the Flipped Learning Scale was based on the existing literature, our research experience regarding teachers' professional development about the integration of learning technologies in classroom settings. An initial list of items was created from previous studies [33, 41, 42]. Following, the various factors were organized thematically, in relation to the research questions, thus creating a new Flipped Learning Scale with 61 Likert-type statements (1 = strongly disagree, 5 = strongly agree) representing teachers' opinions along five dimensions. Table 1 provides sample statements for each dimension of the scale.

Table 1 Examples of statements for each dimension of the scale

Dimension	Item
Pedagogical features of FL	In FL the in-class time is more productive, allowing students to apply new knowledge
Changes in the learning process/students' engagement	I believe that in FL the students are meaningfully involved in problem-solving activities
Teachers' competence in designing and supporting FL	I know what to do as a teacher during the three phases of FL
FL benefits for the students	I believe that in FL students can achieve better learning outcomes
Teachers' challenges and difficulties	I believe that FL requires more time to prepare educational materials and students' learning activities

4 Results

4.1 Descriptive Statistical Analysis

Table 2 summarizes the results of the descriptive statistics regarding the 5 dimensions of the Flipped Learning Scale for teachers. Cronbach's α coefficient was used to measure the internal consistency of the scale dimensions. The values obtained range from 0.79 to 0.96, thus indicating a very high level of internal consistency for the 4 dimensions of the scale [50]. The internal consistency concerning the fifth dimension of the teachers' challenges and difficulties to use flipped learning was high. Therefore, the scale was adequately reliable for measuring teachers' beliefs about flipped learning. The mean levels of the five dimensions in the scale found higher than the mean values of the Likert scale (i.e., 3.0), thus reflecting teachers' positive beliefs about the specific features of the FL model and the factors affecting the implementation of FL intervention sin school practice.

Based on the results in Table 2, it follows that the participant teachers believe that they are sufficiently familiar with the FL model. They appear to recognize the pedagogical features of the FL model ($M = 3.51$) and the changes it brings about in the learning process ($M = 3.60$). In addition, the teachers are positive about the benefits of flipped learning for the students ($M = 3.52$). Most teachers consider that they are competent to design FL interventions ($M = 3.46$), but, on the other hand, they are facing challenges and difficulties to implement FL interventions ($M = 3.62$). The key factors that affect teachers' decisions and attempts to implement FL interventions in their class are related to (a) enhanced workload, (b) the skills need to prepare FL interventions, (c) students' preparation and involvement, and (d) the educational context in primary and secondary schools which is not supportive to teachers' efforts toward integrating flipped learning in their instruction.

Table 2 Descriptive statistics of the scale dimensions

Dimensions	Items	Mean	SD	Cronbach α
Pedagogical features of FL	9	3.51	0.85	0.95
Changes in the learning process/students' engagement	10	3.60	0.75	0.92
Teachers' competencies in designing and supporting FL	12	3.46	0.82	0.95
Benefits for the students	14	3.52	0.83	0.96
Teachers' challenges and difficulties	15	3.62	0.50	0.79

4.2 Gender, Level of Education, Teaching Experience and Specialty

The non-parametric Mann–Whitney U test indicated no statistically significant differences in teachers' opinions across all five dimensions of the flipped classroom scale concerning the factors of gender, school level, and teaching specialty.

In contrast, the Kruskal–Wallis test revealed statistically significant differences regarding the factor of teaching experience across the scale dimensions "Changes in the Learning Process/Students' Engagement" ($H(2) = 6.740$, $p = 0.034$) and "Teachers' Challenges and Difficulties" ($H(2) = 8.449, p = 0.015$). Pairwise comparisons indicated that educators with fewer years of experience (0–10 years) recognize "Changes in the Learning Process" to a greater extent than educators with more than 26 years of experience ($M = 3.68$ and $M = 3.34$, respectively, $p = 0.01$). Furthermore, they rate "Challenges and Difficulties" during the implementation of the FL model higher compared to their colleagues with 11–25 years of teaching experience ($M = 3.78$ and $M = 3.55$, respectively, $p = 0.005$).

4.3 Teachers' Professional Development in ICT

Table 3 presents the results of the five dimensions in the scale concerning teachers' professional development to use ICT in education. The three levels (A, B1, and B2) concern the three consecutive modules of the National In-service Teacher Training Program to effectively use ICT in classroom practice. Pairwise comparisons using the Mann–Whitney U test were conducted to evaluate the hypothesis of different perceptions among teachers who attended the three levels of this professional development program. Statistically significant differences (higher rates) were identified among the teachers who attended the middle B1 level ($U = 7435.0, p = 0.036$) and the advanced B2 level ($U = 7517.5, p = 0.015$), compared to their colleagues who attended only the basic A' level.

The dimension "Teachers' competence in designing and supporting FL" revealed as an important factor that affects their abilities to design and implement effective FL interventions. It is worthy to note that the design of students' learning activities, based on various digital tools and technologies, is a central part of both levels (B1 and B2) in this teacher professional development program. Therefore, the findings confirm existing research regarding the role of teachers' preparation and development of learning design skills.

Table 3 Comparative results (teachers' professional development in ICT)

Dimension	Level A ($N = 113$)			Level B1 ($N = 92$)			Level B2 ($N = 99$)		
	M	SD	*U*	*M*	SD	*U*	*M*	SD	*U*
Pedagogical features of FL	3.59	0.79	9025.5	3.61	0.76	7847.0	3.60	0.82	8087.0
Changes in the learning process/ students' engagement	3.45	0.68	9575.0	3.43	0.73	8823.5	3.49	0.70	8472.0
Teachers' competence in designing and supporting FL	3.46	0.83	9825.5	3.62	0.72	7435.0*	3.62	0.82	7517.5**
Benefits for the students	3.58	0.78	9013.5	3.59	0.78	8088.5	3.60	0.82	8459.5
Teachers' challenges and difficulties	3.62	0.46	9981.5	3.63	0.41	8927.5	3.64	0.44	9091.0

*$p = 0.036$, **$p = 0.015$

4.4 Teachers' Training for Flipped Learning

Table 4 summarizes the results along the five dimensions of the scale concerning the two teacher groups in our analysis; teachers who attended a training program about flipped learning and teachers with no training. The findings of the Mann–Whitney test indicated statistically significant differences between the two teacher groups with respect to the dimensions "Teachers' competence in designing and supporting FL" ($U = 5751.5, p < 0.001$) and "Pedagogical features of flipped learning" ($U = 7944.5$, $p = 0.005$). However, no statistically significant differences were found for the other dimensions in the FL scale.

Table 4 Comparative results (teachers' training for flipped learning)

Dimension	Yes ($N =$ 135)		No ($N =$ 148)		*U*
	M	SD	*M*	SD	
Pedagogical features of FL	3.65	0.78	3.39	0.90	7944.5**
Changes in the learning process/students' engagement	3.47	0.71	3.38	0.79	9244.5
Teachers' competence in designing and supporting FL	3.77	0.68	3.18	0.84	5751.5*
Benefits for the students	3.60	0.76	3.45	0.89	9064.5
Teachers' challenges and difficulties	3.57	0.47	3.69	0.52	11,230.0

*$p < 0.001$, **$p = 0.005$

Table 5 Comparative results (implementation of flipped learning)

Dimension	Implementation of FL				
	No (N = 118)		Yes (N = 165)		
	M	SD	M	SD	U
Pedagogical features of FL	3.18	0.88	3.75	0.75	5727.5*
Changes in the learning process/students' engagement	3.14	0.74	3.63	0.69	5841.0*
Teachers' competence in designing and supporting FL	3.00	0.81	3.79	0.66	4421.5*
Benefits for the students	3.22	0.82	3.74	0.77	5865.5*
Teachers' challenges and difficulties	3.62	0.53	3.62	0.48	10,333.5

*$p < 0.001$

4.5 Implementation of the Flipped Learning Approach

The findings have shown that the experience of implementing flipped learning interventions in their classrooms has a decisive impact on teachers' views, as evidenced by the higher mean values across the four dimensions of the scale (Table 5). The results of the Mann–Whitney test indicated statistically significant differences for the four dimensions of the scale. It is evident that teachers who have applied the flipped learning model in their classrooms have developed a deeper understanding of its pedagogical characteristics, the changes in the processes of students' learning, and the different role the teacher is required to undertake. Furthermore, it appears that they have developed adequate skills in designing flipped learning interventions.

In contrast, no differences were observed regarding "Teachers' challenges and difficulties" associated with adopting the flipped learning model. This dimension is determined by broader factors within the educational context of schools, such as possible deficiencies in technological infrastructure, the increased workload for the teachers in designing flipped learning interventions, the availability of appropriate educational resources, students' attitudes and their preparedness for flipped learning, the overall school culture. It appears that the factors above have the same impact on teachers' views about FL, regardless of the skills they have developed, and, ultimately, play a decisive role in shaping teachers' instructional choices in practice.

5 Conclusions

For purpose of this study, we developed a new scale to measure teachers' views and beliefs about FL in K-12 education. The Flipped Learning Scale is structured along five dimensions: (a) Pedagogical features of FL, (b) Changes in the learning process/ Students' engagement, (c) Teachers' competence in designing and supporting FL, (d) FL benefits for the students and (e) Teachers' challenges and difficulties.

As a general conclusion, it emerges that the teachers in the sample exhibited a positive view of the flipped classroom model, while also recognizing significant challenges and difficulties in implementing flipped classroom interventions. Teachers who have applied the model in their educational practice or attended related training demonstrated a more positive attitude toward the model and appeared to possess the necessary skills for designing FL interventions.

In line with Wang [41] and Yang and Chen [14] the present study showed that teachers' training is an important factor that has a direct influence on their ability to design and implement FL interventions. Our results indicated that the National In-service Teacher Training Program had a positive impact on teachers' beliefs and competence to integrate FL into classroom practice. More specifically, there is a strong correlation with B1 and B2 levels of this program, where the teachers had the opportunity to get involved in designing ICT-based learning scenarios. On the other hand, teachers who have not received any training appeared to possess only marginally sufficient skills and are more hesitant regarding the effectiveness of the flipped learning model.

Overall, the findings clearly indicate that most participants recognized the pedagogical features of the flipped learning (FL) approach and expressed positive views regarding its adoption in school practices and its anticipated benefits for student learning. Additionally, participants considered themselves capable of designing and implementing FL interventions. Consistent with existing research, most teachers acknowledged changes in the learning process, particularly in terms of enhanced student engagement [33, 37, 40]. However, the challenges and difficulties they encountered in implementing FL interventions were associated with their own capabilities, students' attitudes and preparedness, and the broader educational framework in Greek schools.

The findings presented in this paper align with those of a previous qualitative study in Greece, which identified key factors influencing teachers' views on flipped learning across three dimensions [40]: (a) the pedagogical features of the FL model, (b) challenges and concerns related to its implementation, and (c) difficulties in designing FL interventions.

Nevertheless, the effective implementation of flipped learning requires proper student preparation through the pre-class activities, as well as access to suitable digital education resources [8, 38, 41, 42]. Additionally, the design of flipped learning interventions demands significant effort and more time from teachers compared to traditional teaching, especially for preparing educational materials (e.g., video lectures, in-class activities, assessment tools).

Effective implementation of flipped learning requires adequate student preparation through pre-class activities and access to appropriate digital educational resources [8, 38, 39, 41, 42]. Furthermore, designing flipped learning interventions demands considerable effort and significantly more time from teachers compared to traditional teaching, particularly in preparing educational materials such as video lectures, in-class activities, and assessment tools.

The application of flipped learning approaches in Greek schools is not entirely a new concept, in terms of teacher readiness, the pedagogical framework for applying

FL, the role of digital technologies, and teachers' instructional design skills. However, successful implementation requires teachers to make additional efforts by employing specific pedagogical strategies to design and support students' learning in both pre-class and in-class activities. Addressing teachers' concerns, providing targeted professional development, and fostering a collaborative learning environment are critical steps for adopting the FL model and maximizing students' learning outcomes.

The findings of the present study clearly aligned with the classification of teachers' barriers to digital technology integration proposed by Ertmer [51] and Ertmer and Ottenbreit-Leftwich [52]. Specifically, our results identified several first-order barriers to implementing flipped learning in practice, which are external to teachers. These include a lack of adequate educational resources, insufficient time for designing and preparing FL interventions, and limited access to training and support. Second-order barriers pertain to teachers' pedagogical beliefs and competence, such as their understanding of FL pedagogical features, perceived value of FL, their willingness to move away from traditional teacher-centered instruction, and their self-confidence in implementing FL in classroom settings. Consistent with Tsai and Chai [53], third-order barriers are associated with teachers' learning design skills, which they must continuously develop and refine to effectively integrate FL into everyday classroom practices.

Although this study may be limited by the specific sample and the educational context within Greek schools, especially after the COVID-19 pandemic, the findings hold significant value for developing teacher professional development programs and assisting educators in overcoming challenges when designing and implementing FL interventions. Further research combining both quantitative and qualitative data could offer deeper insights into teachers' perceptions and readiness to apply and support flipped learning in regular classroom settings.

In addition, to achieve more in-depth insights regarding teachers' views and beliefs about FL, using the TPACK model could provide a holistic perspective on exploring the integration of FL into everyday educational practice in K-12 schools [54]. Our future efforts will be directed toward combining the FL scale components and the TPACK framework, adapted to the flipped learning pedagogy. It is thus expected to reveal critical factors that affect teachers' competence and readiness to implement flipped learning in their classroom practices.

References

1. Flipped Learning Network: Definition of flipped learning. https://flippedlearning.org/definition-of-flipped-learning (2014). Accessed 1 May 2022
2. Jimoyiannis, A.: Digital Technologies and 21st Century Learning. Kritiki Publishing SA, Athens (2019)
3. Abeysekera, L., Dawson, P.: Motivation and cognitive load in the flipped classroom: definition, rationale and a call for research. Higher Educ. Res. Dev. **34**(1), 1–14 (2015)
4. Tucker, B.: The flipped classroom. Educ. Next. **12**(1), 82–83 (2012)

5. Akçayır, G., Akçayır, M.: The flipped classroom: a review of its advantages and challenges. Comput. Educ. **126**, 334–345 (2018)
6. Baig, M.I., Yadegaridehkordi, E.: Flipped classroom in higher education: a systematic literature review and research challenges. Int. J. Educ. Technol. High. Educ. **20**(1), 61 (2023)
7. Lai, C.L., Hwang, G.J.: A self-regulated flipped classroom approach to improving students' learning performance in a mathematics course. Comput. Educ. **100**, 126–140 (2016)
8. Lo, C.K., Cheung, K.L., Chan, H.R., Chau, C.L.E.: Developing flipped learning resources to support secondary school mathematics teaching during the COVID-19 pandemic. Interact. Learn. Environ. **31**(8), 4787–4805 (2021)
9. Wei, X., Cheng, I.L., Chen, N.S., Yang, X., Liu, Y., Dong, Y., Kinshuk: Effect of the flipped classroom on the mathematics performance of middle school students. Educ. Technol. Res. Dev. **68**, 1461–1484 (2020)
10. Doğan, Y., Batdı, V., Yaşar, M.D.: Effectiveness of flipped classroom practices in teaching of science: a mixed research synthesis. Res. Sci. Technol. Educ. **41**(1), 393–421 (2023)
11. Wang, J., Jou, M., Lv, Y., Huang, C.C.: An investigation on teaching performances of model-based flipping classroom for physics supported by modern teaching technologies. Comput. Hum. Behav. **84**, 36–48 (2018)
12. Linling, Z., Abdullah, R.: The impact of COVID-19 pandemic on flipped classroom for EFL courses: a systematic literature review. SAGE Open. **13**(1), 21582440221148149 (2023)
13. Liu, G.Z., Rahimi, M., Fathi, J.: Flipping writing metacognitive strategies and writing skills in an English as a foreign language collaborative writing context: a mixed-methods study. J. Comput. Assist. Learn. **38**, 1730–1751 (2022)
14. Yang, C., Chen, Y.: Implementing the flipped classroom approach in primary English classrooms in China. Educ. Inf. Technol. **25**, 1217–1235 (2020)
15. Jong, M.S.Y., Chen, G., Tam, V., Chai, C.S.: Adoption of flipped learning in social humanities education: the FIBER experience in secondary schools. Interact. Learn. Environ. **27**(8), 1222–1238 (2019)
16. Birgili, B., Seggie, F.N., Oğuz, E.: The trends and outcomes of flipped learning research between 2012 and 2018: a descriptive content analysis. J. Comput. Educ. **8**(3), 365–394 (2021)
17. De Araujo, Z., Otten, S., Birisci, S.: Mathematics teachers' motivations for, conceptions of, and experiences with flipped instruction. Teach. Teach. Educ. **62**, 60–70 (2017)
18. Jiang, M.Y.C., Jong, M.S.Y., Lau, W.W.F., Chai, C.S., Liu, K.S.X., Park, M.: A scoping review on flipped classroom approach in language education: challenges, implications and an interaction model. Comput. Assist. Lang. Learn. **35**(5–6), 1218–1249 (2022)
19. Strelan, P., Osborn, A., Palmer, E.: The flipped classroom: a meta-analysis of effects on student performance across disciplines and education levels. Educ. Res. Rev. **30**, 100314 (2020)
20. Purwanti, I., Suryawati, E., Eliwarti, E.: Video lectures in online EFL flipped-classroom: effectiveness, students' evaluation and experiences. Eur. J. Educ. Res. **11**(2), 885–898 (2022)
21. Van Alten, D.C., Phielix, C., Janssen, J., Kester, L.: Effects of flipping the classroom on learning outcomes and satisfaction: a meta-analysis. Educ. Res. Rev. **28**, 100281 (2019)
22. Chou, C.L., Hung, M.L., Tsai, C.W., Chang, Y.C.: Developing and validating a scale for measuring teachers' readiness for flipped classrooms in junior high schools. Br. J. Educ. Technol. **51**(4), 1420–1435 (2020)
23. Luo, S., Zou, D.: K-12 teacher readiness for flipped foreign language teaching: scale development and validation. J. Res. Technol. Educ. **56**(6), 674–690 (2024)
24. Bond, M.: Facilitating student engagement through the flipped learning approach in K-12: a systematic review. Comput. Educ. **151**, 103819 (2020)
25. Li, Z., Li, J.: Using the flipped classroom to promote learner engagement for the sustainable development of language skills: a mixed methods study. Sustainability. **14**(10), 5983 (2022)
26. Fischer, I.D., Yang, J.C.: Flipping the flipped class: using online collaboration to enhance EFL students' oral learning skills. Int. J. Educ. Technol. High. Educ. **19**(1), 15 (2022)
27. Giannakos, M.N., Krogstie, J., Sampson, D.: Putting flipped classroom into practice: a comprehensive review of empirical research. In: Digital Technologies: Sustainable Innovations for Improving Teaching and Learning, pp. 27–44. Springer, Cham (2018)

28. Tsai, M.-N., Liao, Y.-F., Chang, Y.-L., Chen, H.-C.: A brainstorming flipped classroom approach for improving students' learning performance, motivation, teacher-student interaction and creativity in a civics education class. Think. Skills Creat. **38**, 100747 (2020)
29. Almodaires, A.A., Alayyar, G.M., Almsaud, T.O., Almutairi, F.M.: The effectiveness of flipped learning: a quasi-experimental study of the perceptions of Kuwaiti pre-service teachers. Int. Educ. Stud. **12**(1), 10–23 (2018)
30. Masood, M., Samaila, K., Chau, K.T.: Application of SQQ-based flipped classroom model on students' achievement and engagement in ICT course. Mediterranean J. Soc. Behav. Res. **6**(1), 21–26 (2022)
31. Merlin-Knoblich, C., Harris, P.N., Mason, E.C.M.: Examining student classroom engagement in flipped and non-flipped counselor education courses. Prof. Counselor. **9**(2), 109–125 (2019)
32. Zainuddin, Z., Haruna, H., Li, X., Zhang, Y., Kai, S., Chu, W.: A systematic review of flipped classroom empirical evidence from different fields: what are the gaps and future trends? Horizon. **27**(2), 72–86 (2019)
33. Kiang, N.H., Yunus, M.: What do Malaysian ESL teachers think about flipped classroom? Int. J. Learn. Teach. Educ. Res. **20**(3), 117–131 (2021)
34. Kaya, Ş., Çebi, A.: Unleashing the potential of flipped learning in K-12: a review of experimental studies. J. Res. Technol. Educ. (2025). https://doi.org/10.1080/15391523.2025.2456052
35. Lin, C.J., Mubarok, H.: Learning analytics for investigating the mind map-guided AI chatbot approach in an EFL flipped speaking classroom. Educ. Technol. Soc. **24**(4), 16–35 (2021)
36. Cheng, L., Ritzhaupt, A.D., Antonenko, P.: Effects of the flipped classroom instructional strategy on students' learning outcomes: a meta-analysis. Educ. Technol. Res. Dev. **67**, 793–824 (2019)
37. Lo, C.K., Lie, C.W., Hew, K.F.: Applying "first principles of instruction" as a design theory of the flipped classroom: findings from a collective study of four secondary school subjects. Comput. Educ. **118**, 150–165 (2018)
38. Yang, C.C.R.: An investigation of the use of the 'flipped classroom' pedagogy in secondary English language classrooms. J. Inf. Technol. Educ. Innov. Pract. **16**, 1–20 (2017)
39. Ye, X.: Insiders' perceptions of the flipped classroom approach in Chinese EFL junior secondary schools. Education. **51**(6), 988–1002 (2022)
40. Dimitrakopoulou, A., Jimoyiannis, A.: Teacher readiness to adopt the flipped learning model: exploring Greek teachers' views and perceptions. In: Reis, A., Barroso, J., Martins, P., Jimoyiannis, A., Huang, R.Y.M., Henriques, R. (eds.) Technology and Innovation in Learning, Teaching and Education, pp. 71–82. Springer, Cham (2022)
41. Wang, T.: Overcoming barriers to 'flip': building teacher's capacity for the adoption of flipped classroom in Hong Kong secondary schools. Res. Pract. Technol. Enhanc. Learn. **12**(1), 6 (2017)
42. Gough, E., DeJong, D., Grundmayer, T., Baron, M.: K-12 teacher perceptions regarding the flipped classroom model for teaching and learning. J. Educ. Technol. Syst. **45**, 390–423 (2017)
43. Jin, Y., Harp, C.Q.: Examining preservice teachers' TPACK, attitudes, self-efficacy, and perceptions of teamwork in a stand-alone educational technology course using flipped classroom or flipped team-based learning pedagogies. J. Digit. Learn. Teach. Educ. **36**(3), 166–184 (2020)
44. Colomo Magaña, A., Colomo Magaña, E., Guillén-Gámez, F.D., Cívico Ariza, A.: Analysis of prospective teachers' perceptions of the flipped classroom as a classroom methodology. Societies. **12**(4), 98 (2022)
45. Sargent, J., Casey, A.: Flipped learning, pedagogy and digital technology: establishing consistent practice to optimise lesson time. Eur. Phys. Educ. Rev. **26**(1), 70–84 (2020)
46. Fisher, R., Perényi, A., Birdthistle, N.: The positive relationship between flipped and blended learning and student engagement, performance and satisfaction. Act. Learn. High. Educ. **22**(2), 97–113 (2021)
47. Toivola, M., Rajala, A., Kumpulainen, K.: Pedagogical rationales of flipped learning in the accounts of Finnish mathematics teachers. Pedagogies. **18**(4), 767–787 (2023)

48. Han, H., Røkenes, F.M., Krumsvik, R.J.: Student teachers' perceptions of flipped classroom in EFL teacher education. Educ. Inf. Technol. **29**(2), 1539–1558 (2024)
49. Jimoyiannis, A., Koukis, N.: Exploring teachers' readiness and beliefs about emergency remote teaching in the midst of the COVID-19 pandemic. Technol. Pedagog. Educ. **32**(2), 205–222 (2023)
50. DeVellis, R.F.: Scale Development: Theory and Applications. Sage, London (2003)
51. Ertmer, P.A.: Teacher pedagogical beliefs: the final frontier in our quest for technology integration? Educ. Technol. Res. Dev. **53**(4), 25–39 (2005)
52. Ertmer, P.A., Ottenbreit-Leftwich, A.: Removing obstacles to the pedagogical changes required by Jonassen's vision of authentic technology-enabled learning. Comput. Educ. **64**, 175–182 (2013)
53. Tsai, C.-C., Chai, C.S.: The "third"-order barrier for technology integration instruction: implications for teacher education. Australas. J. Educ. Technol. **28**(6), 1057–1060 (2013)
54. Roussinos, D., Jimoyiannis, A.: Examining primary education teachers' perceptions of TPACK and the related educational context factors. J. Res. Technol. Educ. **51**(4), 377–397 (2019)

Factors Influencing Teachers' Intention to Use and Acceptance of Big Data in the Greek School Context

Lambros Papoutsakis, Anna Mavroudi, and Spyros Papadakis

Abstract The aim of the present study was to explore the factors influencing primary and secondary education teachers' intention to use Big Data (BD) in their teaching practice. To this end, a quantitative research design was employed, utilizing a structured questionnaire grounded in the theoretical framework of the Technology Acceptance Model (TAM). Analysis of the responses from 750 educators indicated a generally neutral stance and intention toward the integration of BD into their instructional practices. Similarly, their perceptions regarding the perceived usefulness, ease of use, as well as concerns related to data security and privacy, were also characterized by neutrality. Moreover, the findings revealed a strong interest among teachers in receiving professional development and training on Big Data, with the goal of enhancing their ability to utilize such technologies effectively in the educational context. This study constitutes one of the first empirical investigations into the acceptance of Big Data by educators in Greece. Its findings offer valuable insights for both researchers in the field of educational technology and policymakers involved in shaping the integration of emerging technologies within the Greek educational system.

Keywords Big data · School · Teachers · TAM

L. Papoutsakis (✉)
University of West Attica, Athens, Greece
e-mail: lpapoutsakis@uniwa.gr

A. Mavroudi
University of Oslo, Oslo, Norway
e-mail: anna.mavroudi@iped.uio.no

S. Papadakis
Hellenic Open University, Patras, Greece
e-mail: papadakis@eap.gr

I. Kazanidis and A. Tsinakos (eds.), *Research on E-Learning and ICT in Education*,
https://doi.org/10.1007/978-3-031-99961-1_7

1 Introduction

Big Data in education (BDE) is defined as the application of big data techniques to data generated within educational settings, such as students' personal records, audiovisual material, and similar sources [1, 2]. According to [3], the terms 'Big Data in Education' and 'Learning Analytics' are considered synonymous and overlapping. Therefore, in the present study, we adopt the same assumption, without this being regarded as a conceptual or methodological error. Big Data is already being utilized in a wide range of fields, including healthcare, transportation, and finance [4]. Education is among these sectors, with a growing number of researchers and countries showing increasing interest in the use of BD in educational contexts [5].

The advantages of employing BD in education are primarily associated with improvements in two key areas: (a) the administrative operations of educational institutions, and (b) the teaching and learning process [6]. Specifically, BD can support schools and educational institutions in enhancing the efficiency of administrative services and procedures related to logistics, maintenance, and resource management [7]. Additionally, BD can contribute to improving student performance and learning outcomes by enabling more personalized and adaptive learning experiences [8]. Another notable benefit is the ability of BD to assist students in making informed decisions about their academic and professional futures based on their performance data [9].

Moreover, the algorithms leveraged by BD in education can predict student achievement, thereby supporting the refinement of teaching practices and helping to identify and address potential weaknesses within the educational system [10].

Despite its growing relevance, the issue of BD acceptance in Greek schools has not been sufficiently explored. According to the existing literature, only two studies have thus far examined the acceptance of BD in primary and secondary education settings [11, 12]. Understanding user acceptance is crucial, as the successful adoption of any new technology depends on its acceptance by its intended users. In the context of individual-level technology acceptance, numerous studies have been conducted, utilizing a variety of models and theories to predict and explain human behavior regarding technology adoption [13].

It is well established that for a technology to be accepted by its potential users, it must first be perceived as new and innovative [14, 15]. In the educational domain, teachers are among the most influential stakeholders. It is therefore deemed essential to investigate the acceptance of BD by primary and secondary school educators, as well as the methods by which BD technologies are employed, especially given the promising findings associated with their implementation in educational settings.

2 Methods

2.1 *Purpose*

The present study seeks to address the research gap concerning the acceptance and use of BD in schools. It aims to answer the following research questions:

RQ1: What factors may influence the intention to use Big Data in schools?

RQ2: What are the factors that could facilitate the use of Big Data in Greek schools by teachers?

RQ3: What are the preferred ways in which teachers envision using Big Data in the school context?

2.2 *Literature Review on the Acceptance of Big Data in Education*

Regarding the acceptance of practices and methods for analyzing Big Data (BD) within the Greek school system, [11] conducted a study examining how Greek educators might integrate BD into educational settings. This case study involved 30 teachers who responded to questions concerning: (1) whether student data is digitally stored at their schools, (2) what types of data are collected and how these data are utilized within the educational process, and (3) which online tools are employed and in what ways. The findings indicated that such data are primarily used to fulfill administrative needs within the school unit, as well as to support the teaching and learning process.

Building on this initial research, [10] conducted a quantitative study involving both educators and educational administrators, aiming to assess the acceptance of BD analysis and its application in educational practice. This study was grounded in the Technology Acceptance Model (TAM) [14], which is further elaborated below. The responses of 98 participants revealed a generally positive attitude toward the use of BD as a pedagogical tool. Specifically, participants indicated that BD could support communication with parents or guardians regarding student progress, as well as assist in identifying students at risk of academic failure. Furthermore, the use of such technologies was perceived as a means of enhancing teachers' professional profiles.

The research conducted by [16] examines the key determinants that shape teachers' engagement with digital data in upper secondary education in Switzerland, focusing on the use of learning analytics. Utilizing a large-scale survey involving 1059 teachers across 54 schools, the study assessed various factors such as teachers' attitudes, perceived competence in data use, technological infrastructure, and contextual elements like the frequency of student device usage. The findings validated the applicability of the Will-Skill-Tool (WST) model, identifying teachers' attitudes (will), their data literacy (skill), and access to relevant technologies (tool) as significant predictors of digital data utilization. Among these, data literacy was found to be the most influential individual factor. Other elements, including students' digital habits, educators' gender, and teaching experience, showed minor yet notable effects, whereas the frequency of teachers' own device use did not significantly impact data practices. Despite the availability of analytics-equipped platforms, overall reported use and literacy remained low. These results underscore the need for tailored professional development and call for deeper insights into how educational platforms and institutional environments influence data-driven teaching.

The empirical study of [17] explores the determinants influencing higher education educators' intention to adopt big data technologies in teaching, using the Unified Theory of Acceptance and Use of Technology (UTAUT) as a theoretical framework. A quantitative survey was administered to 193 educators in Yunnan Province, China, and the data were analyzed using structural equation modeling (SEM) to evaluate the relationships among the UTAUT constructs. The results indicated that performance expectancy and facilitating conditions significantly and positively affect educators' behavioral intention to integrate big data technologies into their teaching practices. In contrast, effort expectancy and social influence did not show statistically significant effects. Additionally, demographic factors such as gender, age, and teaching experience did not account for meaningful differences in behavioral intention. These findings emphasize the importance of enhancing perceived usefulness and institutional support to encourage technology adoption in academia.

The findings of these studies highlight a significant research gap concerning the investigation of BD acceptance and use across all levels of education.

2.3 Theoretical Framework

Researchers in the field of educational technology have drawn upon a range of theories and acceptance models to explain the adoption and use of technology. One of the most prominent and widely applied models is the Technology Acceptance Model (TAM) [14]. According to the meta-analysis conducted by [18], TAM is considered a valid and reliable framework for explaining technology acceptance among educators.

TAM suggests that an individual's intention to use a technology serves as a key predictor of actual usage. The core constructs of the model include intention, attitude, perceived usefulness, and perceived ease of use. Researchers often extend the model by incorporating additional variables that are contextually relevant to the specific domain in which technology acceptance is being examined.

Within the TAM framework, perceived ease of use influences both perceived usefulness and attitude toward use. In turn, perceived usefulness and attitude shape the intention to use the technology, which subsequently affects the actual usage behavior.

2.4 Research Methodology

Data were collected through a structured questionnaire designed for the purposes of a quantitative study. The questionnaire was based on the theoretical framework of the Technology Acceptance Model (TAM) [14] and was extended by incorporating the variable of perceived risk related to the security and privacy of student data, in response to concerns documented in the literature such as [19, 20].

The instrument consisted of three sections and included 32 closed-ended questions. Prior to completing the questionnaire, participants were asked to watch an introductory video on the topic of Big Data in education (BDE).

The first section contained six items related to the demographic and professional characteristics of the participants (e.g., teaching discipline, years of service).

The second section included 21 items measured using a five-point Likert scale (ranging from 1 = not at all to 5 = very much). These questions assessed teachers' attitudes and perceptions regarding the use of BD in schools and their willingness to receive professional training on the subject. The items were derived from:

(a) The core constructs of the TAM (i.e., intention to use, attitude, perceived usefulness, and perceived ease of use), and.
(b) Validated items from [20] addressing the perceived risk associated with data security and privacy.

The third section comprised five additional Likert-scale items focusing on potential ways to utilize BD in schools. These questions were informed by the study of [12], as well as research into facilitating factors for the adoption of other innovative educational technologies, such as smart glasses [21].

The questionnaire was distributed electronically, using the email networks of primary and secondary school units. In total, 750 educators from primary and secondary education responded.

The statistical analysis of the data was limited to basic descriptive statistics, including mean, median, and standard deviation.

3 Results and Discussion

3.1 Demographic and Professional Characteristics of Participants

A total of 750 primary and secondary education teachers participated in the study. Of these, 501 (66.8%) identified as female, 247 (32.9%) as male, while two participants (0.3%) did not disclose their gender.

In terms of age, 80 participants (10.7%) were between 22 and 35 years old, 173 participants (23.1%) were between 36 and 45 years old, and 497 participants (66.3%) were over the age of 46.

Table 1 Number of items and reliability of variables

Variable	No of questions	Cronbach A
Intention to use	3	0.943
Perceived usefulness	6	0.960
Perceived ease of use	5	0.786
Perceived risk	6	0.792
Facilitating conditions	5	0.837
Ways of use	4	0.868

With regard to the educational level in which they worked, 314 (41.9%) were employed in primary education, and 436 (58.1%) in secondary education.

Participants held the following professional roles:

(a) 12 were school advisors/education coordinators (1.6%),
(b) 195 were school principals or vice-principals (26.0%), and
(c) 543 were classroom teachers (72.4%) across a wide range of subject areas and specialties.

Concerning teaching experience, 165 participants (22.0%) had 1–10 years of service, 222 (29.6%) had 11 to 20 years, and 363 (48.4%) had more than 20 years of teaching experience.

3.2 Reliability of the Study

Table 1 presents the results of the reliability analysis for each of the variables, as measured by Cronbach's alpha coefficient. All variables yielded values greater than 0.7, which is considered satisfactory for ensuring reliable measurements [22].

3.3 Acceptance of Big Data in Schools by the Surveyed Educators

Table 2 presents the descriptive statistics for the variables of intention to use, perceived usefulness, perceived ease of use, and perceived risk associated with BD. Overall, the results indicate a neutral stance among educators with respect to all four variables.

For each item, the mean and median values are closely aligned, which reinforces this interpretation. It is noteworthy that perceived ease of use scored slightly lower compared to the other three variables. Additionally, the data reveal a slightly negative attitude among educators regarding the safeguarding of data security and privacy.

Regarding participants' desire to receive training in BD (LA), the descriptive statistics indicate a strong inclination (mean = 3.67, median = 3.19, SD = 1.33).

Table 2 Mean, median, and standard deviation of the variables

Variables	Mean	Median	S.D
Intention	3.21	3.22	1.18
I intend to use BD in my future teaching	3.22	3.26	1.12
I plan to use BD in my future teaching	3.07	3.11	1.25
I anticipate that I will use BD in my future teaching	3.33	3.42	1.18
Perceived usefulness	3.35	3.33	1.18
Using BD in my teaching would allow me to complete teaching-related tasks more quickly	3.39	3.5	1.17
Using BD in teaching would improve my performance	3.31	3.37	1.18
Using BD in teaching would enhance my productivity	3.28	3.32	1.22
Using BD in teaching would increase my effectiveness	3.33	3.43	1.17
Using BD in my teaching would make it easier	3.33	3.41	1.18
I find that using BD in my teaching would be useful	3.48	3.67	1.14
Perceived ease of use	3.08	3.13	1.11
Using BD would require significant effort on my part	3.24	3.24	1.20
I find it difficult to use BD in my work	2.75	2.93	1.25
It is easy for me to understand how to use BD in my work	3.19	3.22	1.09
I find it clear and understandable how to use BD in my work	3.11	3.11	1.09
Overall, I believe that BD will be easy to use	3.13	3.15	0.92
Perceived security and privacy risk	3.18	3.30	1.26
BD collect excessive amounts of user information	3.27	3.27	0.95
BD collect excessive amounts of information about my students	3.40	3.41	0.96
I would be concerned about my students' privacy when their data are processed through BD systems at school	3.34	3.39	1.34
I have doubts about how well my students' privacy is protected when their data are processed through BD systems at school	3.33	3.4	1.43
Unknown third parties might be able to access my students' data when I use BD in my teaching	2.54	3.39	1.65
My students' data are securely stored within the structures and digital systems of the Ministry of Education and Religious Affairs, e.g., the Panhellenic school network	3.21	3.21	1.22

Table 3 presents the descriptive statistics concerning the factors and conditions that could potentially facilitate the future use of BD in teaching. The most influential factor identified was access to the Internet, while the least influential was the imposition of mandatory use by the Ministry of Education and Religious Affairs.

Table 3 Mean, median, and standard deviation of the facilitating factors

Facilitating factors and conditions	Mean	Median	S.D.
Internet access in the classroom	3.97	4.3	1.37
Training in the management of BD (technological aspects)	3.92	4.16	1.29
Support from school leadership regarding the management of BD	3.79	4.2	1.28
Mandates from official educational policy through legislation	2.80	2.89	1.69
Support from school advisors/coordinators of educational work	3.43	3.57	1.43

Table 4 Teachers' preferences regarding the uses of learning analytics

Preferences regarding the uses of learning analytics	Mean	Median	S.D
Real-time feedback on student performance	3.80	4.2	0.97
Prediction of students' risk of academic failure	3.48	3.55	1.13
As a recommendation tool for selecting appropriate learning activities for each student	3.79	4.12	0.92
As a recommendation tool for the effective formation of student work groups	3.74	4.05	0.91

3.4 Potential Uses of Learning Analytics in Schools

Table 4 presents the descriptive statistics related to teachers' preferences regarding four potential uses of BD in schools:

(a) real-time feedback on student performance,
(b) prediction of students' risk of academic failure,
(c) as a recommendation tool for selecting appropriate learning activities for each student, and,
(d) as a recommendation tool for the effective formation of student work groups.

Overall, respondents appear to hold a positive attitude toward these proposed uses. More specifically, they seem to express greater interest in real-time feedback on student performance and less interest in the use of BD for predicting students' risk of failure.

4 Conclusions

This study aimed to examine the factors influencing the intention of primary and secondary school teachers to use BD, the conditions that facilitate their implementation in school settings, as well as teachers' preferred modes of use.

Regarding the first research question—namely the factors influencing the intention to use BD—it was found that teachers generally held a neutral stance toward the use of BD in their teaching. Similarly, teachers' perceptions of the usefulness and ease of use of BD in their instructional practice were also neutral. The same applied to their views on student data security and privacy. This neutrality may be attributed to a lack of technical knowledge about BD technologies and their potential benefits. Further research is needed on the acceptance of BD by teachers, a need also highlighted by [11]. The same can be said concerning the necessity for professional development, as evidenced by the findings of [8].

With respect to the second research question—the factors that could facilitate the use of BD—teachers reported that implementation would be more feasible if: (a) there were Internet access in their classrooms, (b) they received training on technical aspects of BD, and (c) they were supported by school leadership and educational coordinators. This finding—i.e., the importance of specific facilitating factors—aligns with previous studies on the acceptance of other innovative technologies in the field of education (e.g. [21]).

Regarding the third research question—teachers' preferred uses of BD—it was found that educators would be most interested in using BD for: (a) real-time feedback on student performance, (b) recommending appropriate learning activities for each student, (c) supporting the effective formation of student work groups, and (d) predicting students' risk of academic failure. These findings are consistent with those reported by [12].

This study contributes both to research and educational practice. Specifically, it enriches the academic literature on the acceptance of BD in education, with an emphasis on the school context. Its findings may serve as a foundation for further investigation by BD specialists, who may explore either the acceptance of such systems or the sources of data extraction within the context of centralized educational systems. A centralized educational system is one in which a central authority, typically a national government or ministry of education, holds primary control over key aspects of education. This includes decision-making related to curriculum design, resource allocation, teacher recruitment, and the establishment of educational policies. It offers uniformity and centralized control, which can lead to consistency and equity across the educational landscape. However, it may also limit local flexibility and responsiveness to specific community needs [23–25]. Furthermore, this research provides valuable insights for the Ministry of Education and Religious Affairs, particularly in terms of integrating BD into official policy decisions. One such initiative could be the inclusion of BD in the national professional development program for teacher training. Additional training should also be provided to teachers on technical issues such as the General Data Protection Regulation (GDPR), pedagogical approaches, and state-mandated digital data use policies.

5 Limitations and Future Research

This study has three main limitations. First, the sample used was one of convenience. Second, the study examines the acceptance of BD within the context of the Greek educational system, which is characterized as centralized [23]. Therefore, the generalization of the findings to samples from other countries with different educational systems should be approached with caution. Third, the teachers in the sample were not sufficiently familiar with the concept and use of BD. This implies that their involvement in activities directly related to the use of BD might have produced different results.

Future research could expand upon this study by applying the questionnaire to a representative sample of teachers. Additionally, variables from other acceptance models and theories, such as social influence and self-efficacy, could be incorporated, followed by inferential statistical analysis to explore their impact.

References

1. Daniel, B.K.: Big data and data science: a critical review of issues for educational research. Br. J. Educ. Technol. **50**(1), 101–113 (2019)
2. Poulovassilis, A.: Big Data and Education, pp. 1–23. Birkbeck Knowledge Lab, London (2016)
3. Romero, C., Ventura, S.: Educational data mining and learning analytics: an updated survey. WIREs Data Min. Knowl. Discov. **10**(3), 1355 (2020). https://doi.org/10.1002/widm.1355
4. Zhang, H., Lee, S., Lu, Y., Yu, X., Lu, H.: A survey on big data technologies and their applications to the metaverse: past, current and future. Mathematics. **11**(1), 96 (2022)
5. Fischer, C., et al.: Mining big data in education: affordances and challenges. Rev. Res. Educ. **44**(1), 130–160 (2020)
6. Daniel, B.: Big data and analytics in higher education: opportunities and challenges. Br. J. Educ. Technol. **46**(5), 904–920 (2015)
7. Kamal, J., Dave, M.: A framework for managing and analyzing big data in Indian school education system with reference to Jammu & Kashmir. Int. J. Innov. Technol. Explor. Eng. IJITEE. **8**, 885–890 (2019)
8. Drigas, A.S., Leliopoulos, P.: The use of big data in education. Int. J. Comput. Sci. Issues IJCSI. **11**(5), 58 (2014)
9. Yang, X., Ge, J.: Predicting student learning effectiveness in higher education based on big data analysis. Mob. Inf. Syst. **2022**(1), 8409780 (2022)
10. Wang, P., Zhao, P., Li, Y.: Design of education information platform on education big data visualization. Wirel. Commun. Mob. Comput. **2022**(1), 6779105 (2022)
11. Mavroudi, A., Papadakis, S.: A case study on how Greek teachers make use of big data analytics in K-12 education. In: International Symposium on Emerging Technologies for Education, pp. 3–9. Springer, Cham (2019)
12. Mavroudi, A., Papadakis, S., Ioannou, I.: Teachers' views regarding learning analytics usage based on the technology acceptance model. TechTrends. **65**(3), 278–287 (2021)
13. Granić, A.: Technology adoption at individual level: toward an integrated overview. Univ. Access Inf. Soc. **23**(2), 843–858 (2024)
14. Davis, F.D.: Perceived usefulness, perceived ease of use, and user acceptance of information technology. MIS Q., 319–340 (1989)
15. Rogers, E.M., Singhal, A., Quinlan, M.M.: Diffusion of innovations. In: An integrated approach to communication theory and research, pp. 432–448. Routledge, London (2014)
16. Michos, K., Schmitz, M.-L., Petko, D.: Teachers' data literacy for learning analytics: a central predictor for digital data use in upper secondary schools. Educ. Inf. Technol. **28**(11), 14453–14471 (2023)
17. Wang, Q., Jalil, H., Marof, A.M.: Factors affecting the acceptance of big data technology in teaching among higher education educators: an empirical investigation using the UTAUT model. Int. J. Acad. Res. Bus. Soc. Sci. **12**(12), 1049–1066 (2022)

18. Scherer, R., Siddiq, F., Tondeur, J.: The technology acceptance model (TAM): a meta-analytic structural equation modeling approach to explaining teachers' adoption of digital technology in education. Comput. Educ. **128**, 13–35 (2019)
19. Wang, Y.: Big opportunities and big concerns of big data in education. TechTrends. **60**, 381–384 (2016)
20. Al-Rahmi, W.M., et al.: Big data adoption and knowledge management sharing: an empirical investigation on their adoption and sustainability as a purpose of education. IEEE Access. **7**, 47245–47258 (2019)
21. Kazakou, G., Koutromanos, G.: Augmented reality smart glasses in education: teachers' perceptions regarding the factors that influence their use in the classroom. In: Interactive Mobile Communication, Technologies and Learning, pp. 145–155. Springer, Cham (2021)
22. De Vellis, R.: Scale Development: Theory and Applications, 2nd edn. Sage Publications, Thousand Oaks (2003)
23. Koutouzis, M.: Educational organizations and the Greek educational system. In: Athanasoula-Reppa, A., Dakopoulou, A., Koutouzis, M., Mavrogiorgos, G., Chalkiotis, D. (eds.) Educational unit management, 2nd edn, pp. 40–44. Hellenic Open University, Patras (2008)
24. Eteokleous, N.: Evaluating computer technology integration in a centralized school system. Comput. Educ. **51**(2), 669–686 (2008)
25. Brown, S., McIntyre, D.: An action-research approach to innovation in centralized educational systems. Eur. J. Sci. Educ. **3**(3), 243–258 (1981)

Instructional Design, Artificial Intelligence, and Serious Games

Directed AI: The Future of AI for Instructional Design

George Palaigeorgiou, Panagiota Christodoulou, and Ioannis Kazanidis

Abstract The integration of generative AI in education is transforming instructors and instructional designers into directors of AI-driven learning experiences. This article explores how AI empowers educators to transcend traditional content creation by curating, orchestrating, and guiding dynamic, personalized educational journeys. Instructors are expected to direct learning experiences with a clear vision, aggressively explore the accessible and expansive design space, and craft engaging narratives that connect emotionally and intellectually with learners. They also play a pivotal role in providing meta-level instructional designs that enable individualized learning pathways and creating non-linear, multimodal experiences tailored to diverse learner needs. What was once a block-by-block design process is now a dynamic practice of rapidly prototyping elastic, adaptive learning scenarios. The barriers that once constrained creativity—logistical challenges, cognitive biases, and time limitations—are now less with AI. By adopting the role of directors, instructors must strategically blend human insight with AI capabilities, balancing innovation with ethical responsibility.

Keywords Generative AI · Directed AI · Instructional design · Instructors

G. Palaigeorgiou (✉) · P. Christodoulou
LearnWorlds Ltd, Limassol, Cyprus
e-mail: gpalegeo@learnworlds.com

P. Christodoulou
e-mail: panagiota@learnworlds.com

I. Kazanidis
Department of Informatics, Democritus University of Thrace, Kavala, Greece
e-mail: kazanidis@cs.duth.gr

I. Kazanidis and A. Tsinakos (eds.), *Research on E-Learning and ICT in Education*,
https://doi.org/10.1007/978-3-031-99961-1_8

1 Introduction

Throughout the history of educational technology, we can identify repeated cycles of unfulfilled promises with each new development. From the early use of instructional television and radio to the widely accessible internet, computer-assisted instruction, and MOOCs, each innovation was advocated as a transformative force capable of revolutionizing learning. These technologies have never accomplished what they promised [1–4] and the main reason for this is their passive nature and focus on content delivery over critical thinking and engagement. Selwyn [4] states that this shortcoming of the educational technologies is fundamental and relates to the material and systemic structures inherent in shaping the educational technology. Mainstream for too long, learning technologies have been built around institutional values—efficiency, standardization, and control—disenfranchising values like autonomy, creativity, and care. This has given rise to what is commonly referred to as an "illusion of learning" which these technologies tend to perpetuate, but lacks the deep cognitive engagement [5]. These repeated patterns reveal a deeper, more systemic problem: true learning requires active cognitive effort, emotional resilience, metacognitive awareness, and self-regulation.

In this sense, generative AI is a chance for most researchers and media alike to move away from the limitations of its predecessors. In contrast with the earlier technologies, generative AI focuses on personalization, active involvement and personalized adaptive learning routes based on the cognitive and affective needs of individual learners. Generative AI allows for automatic, high-quality feedback to students (e.g., [6–8]), interactive and multimodal tools (e.g., [9, 10]), dynamic content integration (e.g., [10]), and improved learning outcomes (e.g., [11, 12]), and transforms educators from content creators to mentors and designers of personalized learning experiences.

AI works by linking data, ideas and concepts from various sources. These connections are similar to how our brains join pieces of information, which enables an AI to merge knowledge from multiple disciplines to create new concepts that would otherwise be unconnected. AI is powerful because on the one hand it keeps a balance between strong connections that help it recall accurate information and on the other side loose connections that allow AI to think outside the box.

AI uses "*strong connections*" in order to generate results that are accurate and precise. This is possible because AI processes large amounts of information quickly, identifies patterns, understands context, and provides useful insights. Additionally, by encoding centuries of collective knowledge, AI enables educators to exploit and synthesize interdisciplinary resources and integrate them into their teaching practices.

Nonetheless, AI also takes advantage of "*loose connections*" in order to formulate creative and surprising links between different ideas, thus generating creative results. To illustrate, AI can combine things that are usually unrelated, like relating ancient myths with science fiction or blending cubism with pop art, ensuring more unexpected results that challenge traditional boundaries. Unlike instructors, who are often constrained by cognitive biases and cultural norms like the "anchoring" effect, where our first idea influences all subsequent thinking, AI operates free from these limitations and presents outputs that push instructors out of their comfort zones. This kind of "out-of-the-box" thinking is difficult for humans to replicate at scale.

In this article, we will explore how instructors and instructional designers can move beyond their traditional roles to become directors of AI-driven learning experiences. Instructors and instructional designers will need to adopt new frameworks and meta-level design strategies in order to employ AI's generative and adaptive capabilities to deliver transformative, impactful, learner-centered education.

2 Instructors and AI

2.1 AI Frameworks for Instructors

As AI becomes more common in education, the role of instructors and instructional designers is changing significantly. To help with this shift, various stakeholders and researchers developed frameworks (see [13] for a review) to guide educators on the competencies, skills, and knowledge needed to exploit AI effectively. These frameworks focus on giving educators the tools to adjust to the changes AI is bringing to education. The majority of these frameworks help teachers understand the technical side of AI as well as stress the importance of ethics, flexible teaching methods, and creating environments that encourage collaboration.

A holistic framework was suggested by UNESCO, the AI Competency Framework for Teachers, emphasizing foundational AI literacy and ethical awareness. It stresses pedagogical adaptation, a human-centered mindset, ethical use, technical AI knowledge about concepts and tools, and AI for professional development [14]. The framework suggests three levels of progression in AI competencies (i.e., Acquire, Deepen, Create). These progression levels allow educators to identify their current AI competencies and set their learning objectives aiming for professional development [14]. Therefore, one can argue that this framework encourages stakeholders' efforts to include AI education in teacher training programs. It also emphasizes the importance of changing the role of teachers, moving from being just content creators to becoming directors can guide the design of AI-powered, student-centered learning experiences.

The Intelligent-TPACK Framework builds on the Technological Pedagogical Content Knowledge (TPACK) model by adding AI to improve educators' understanding of technology, teaching methods, and subject matter [15]. It focuses on using AI to support personalized learning, provide adaptive feedback, and develop new teaching strategies while also helping teachers gain a strong understanding of how AI tools work and can be applied across different subjects. The framework also stresses the ethical use of AI, tackling issues like transparency, bias, and data privacy, and encourages an interdisciplinary approach to designing curriculum and teaching.

Filo et al. [16] suggested the AI Competency Framework for Teachers and Students, which stands out for its co-creation with educators approach, ensuring practical relevance. For teachers, the framework focuses on using AI to improve personalized learning, flexible teaching, and innovative instruction. Additionally, it promotes the idea of AI agency, namely the skill of a critical and intentional approach to using AI aligned with ethical considerations. The framework also highlights the need for teachers to understand how AI works while developing critical attitudes and ethical values to guide the effective use of AI in education.

Another AI framework emphasizing the level of educators' readiness to implement AI in learning and instruction was suggested by Luckin et al. [17], offering a 7-step EThICAL AI Readiness framework (i.e., Excite, Tailor & Hone, Identify, Collect, Apply, Learn, Iterate) to educators implementing AI responsibly and effectively. The framework empowers educators by highlighting that AI readiness is an active process that needs a participatory training approach that will eventually empower educators to be able to exploit AI to meet their needs.

Further, a framework attempting to identify the AI digital competencies in the post-pandemic world was suggested by Ng et al. [18]. This framework adapts existing digital competency models (e.g., DigCompEdu, P21 Framework for twenty-first Century Learning) to reflect AI-specific needs in teaching and learning. It highlights four domains: teacher professional engagement to enhance teaching and stakeholder collaboration; instructional design to align pedagogies, technologies, and assessments with student learning goals; content choices to foster AI literacy across disciplines, from basic knowledge to advanced ethical and creative skills; learning competencies to develop student's knowledge, skills, and values for AI readiness, emphasizing life, innovation, and technological skills.

Finally, the five layers of expertise in AI and instructional design framework by Palaigeorgiou and Christodoulou [19] focuses on instructional designers, highlighting five layers of AI expertise. These layers refer to how AI can facilitate instructional designers to understand the domain where they need to design a learning experience, produce learning activities and instructional materials, ensure efficiency in the design of instructional plans through prompt engineering, follow instructional design theories as well as strategies of creative and critical thinking for the design of instructional plans, and mastering AI tools. The added value that this framework brings to the literature is that it underlines the leading role of designers in orchestrating AI-powered learning experiences.

While the previous frameworks contribute significantly to the field, they collectively underscore the need for a paradigm shift in the perception of the educator's role without envisioning this role.

2.2 Instructors' Role

Although the literature about the exploitation of AI in education is flourishing, there is a limited number of studies discussing how AI can benefit the process of instructional design and the role of the instructor (e.g., [20–23]). However, the landscape is rapidly evolving, with AI-powered tools already making significant strides in transforming the instructional design process. For instance, platforms like LearnWorlds (www.learnworlds.com) use AI to produce quickly high-quality, creative, multimodal educational materials, including course outlines, eBooks, assessments, and interactive videos, rendering content creation more efficient than in the past.

Most studies highlight that AI has the potential to *automate the labor-intensive processes* involved in designing and developing instructional materials [24–26]. This enables instructors and instructional designers to shift their focus to higher-value tasks, such as customizing lessons to meet individual student needs or creating innovative, engaging learning experiences rather than spending time on material production [27, 28]. AI also helps educators easily adapt and reuse content in different formats. For example, it can turn text-based lessons into podcasts or video tutorials. This makes learning materials more accessible and allows educators to reach a broader audience.

Beyond efficiency, recent studies [27, 28] emphasize the evolving role of instructors and instructional designers as *curators of quality educational content* in an era of abundant AI-generated resources. While AI can create lesson plans, essays, and other materials, teachers still need to carefully review these outputs to ensure they are accurate, relevant, and aligned with learning goals [26]. For example, AI-generated content should be checked for mistakes, biases, or anything unfitting with the curriculum. Therefore, educators act as gatekeepers, improving and personalizing AI-generated materials to create meaningful and customized learning experiences for their students. By working with AI tools, educators can enhance content while using their professional judgment to adapt it to the unique needs of their students.

Studies have also attributed to AI a significant *shift to personalized and multimodal instructional approaches*, enabling adaptable and engaging educational experiences (e.g., [26, 29, 30]). For instance, a recent meta-analysis found that incorporating AI-powered virtual agents into computer-based simulations for learning yielded a medium positive overall effect compared to conventional computer-based simulations alone [31]. Another study highlighted the benefits of multimodal learning through an AI-powered system that utilized text-to-text and text-to-image LLMs to deliver real-time, customized educational content. This system adapted to students' language learning skills through an engaging, visually grounded game, played either with or without a social robot [32]. The study's results revealed that both groups had significant vocabulary gains, regardless of the presence of the social robot.

AI's integration into education also shifts the role of instructors toward *being strategic decision-makers and facilitators.* Particularly, predictive and learning analytics provide detailed insights into student performance by identifying strengths and weaknesses (e.g., [33–35]). Still, the instructor remains central in analyzing AI-generated insights and making informed decisions about instructional methods, interventions, or the delivery of personalized support. Beyond technical insights, instructors bring empathy, encouragement, and tailored guidance to effectively address challenges, emphasizing that education is not only about delivering information but also about raising resilience and personal growth. This evolving role of instructors exemplifies the concept of Human–AI Hybrid Intelligence Systems, where human expertise and AI capabilities are combined to maximize educational outcomes [36].

For instance, an instructor might use AI analytics to spot where a learner face difficulties, while, at the same time, rely on their expertise to give meaningful feedback and encouragement to the learner. Using this approach the strengths of AI are combined with human judgment.

The use of AI in teaching also highlights the critical role of educators in helping students understand the *ethical use of AI* and its impact on society [37]. As AI becomes more common, it's vital for students to learn about ethical issues like algorithmic bias, data privacy, and responsible use [28]. Teachers can lead conversations about how AI affects the environment and society, encouraging students to think critically about its benefits, challenges, and limitations. Such discussions align with the principles of Responsible Research and Innovation (RRI), which advocate for inclusivity, anticipation, reflexivity, and responsiveness in the development and use of emerging technologies [38]. Educators who will include RRI into their teaching can present AI as a societal force requiring careful and responsible application [39].

3 Why Instructional Designers and Instructors Will Become Directors

Traditionally, the most significant challenge for instructors was developing an instructional plan and crafting the corresponding materials and activities—a process that required weeks or months of effort. AI has turned this paradigm on its head, enabling the production, testing, and refinement of high-quality content in hours. This transformation doesn't just solve old problems—it creates new ones centered around managing the enormous design space AI opens up. Instructors are no longer limited by time or resources but are instead faced with the task of curating, orchestrating, and directing vast creative possibilities into purposeful and impactful learning experiences. In this context, the role of the instructor evolves from a creator of individual lessons to a *director of AI-driven learning experiences.*

3.1 Directing Creativity Toward a Vision

AI has democratized creativity in instructional design allowing instructors to design anything they can imagine [25]. However, this vast range of possibilities means educators need to shift their focus from just creating content to thinking strategically about how they use their creativity during the design process. Specifically, they need to turn their ideas into clear, meaningful goals to ensure their work has purpose and direction. The instructors must align every decision with an overarching vision, considering their learning objectives, the cognitive outcomes and learners' emotional engagement. Like a director overseeing a film, the instructors must align every creative decision with a unified vision, choosing which elements to emphasize and how to combine them for the greatest educational impact.

3.2 Opening and Navigating in the Instructional Design Space

AI revolutionizes instructional design by enabling a breadth-first exploration of the design space, shifting away from the traditional depth-first approach. Till now, instructors focused on upgrading a single "perfect" lesson plan through iterative improvements. Now, AI supports educators and instructional designers in producing and exploring multiple variations of instructional materials or approaches simultaneously, thereby significantly expanding the design space. Shifting from the depth of one design to the breadth of multiple designs creates opportunities for creativity and innovation, allowing different and potentially innovative solutions to emerge in educational design. In addition, rapid prototyping, is supported allowing instructors to identify what works best and for which categories of learners, thus promoting adaptability by uncovering solutions tailored to varied learning needs.

Additionally, this approach helps educators make better decisions by allowing them to compare different design options side by side, lowers risks by offering a variety of choices and boosts creativity by focusing on exploring new ideas rather than rushing to perfect a single one.

Nevertheless, the expansion of the design space holds a key challenge for instructors, namely the development of a new set of skills, dispositions and competences. Particularly, instructors and instructional designers need to learn how to use AI tools not just to generate multiple designs, but to ask the right questions, shape creative possibilities, and explore ideas that align with their teaching goals and overall educational vision. Thus, instructors need a new level of strategic thinking and the disposition of openness to experimentation. Instead of refining a single plan, instructors now focus on exploring multiple design solutions, selecting the most effective ones, and combining them to create engaging and impactful learning experiences. This shift represents a significant change, as the instructor's role expands to include curation, synthesis, and exploration.

3.3 Directing Authentic and Immersive Learning Narratives

AI empowers instructors to transcend their traditional roles as content creators and step into the transformative role of storytellers. Previous research underscores the numerous benefits of storytelling, including increased student engagement and motivation, deeper reflection, and the development of critical twenty-first-century literacies [40, 41]. Stories and narratives connect ideas to real-life situations, emotions, and cultural contexts, making learning feel more authentic and meaningful. Therefore, creating stories in the previous context is in line with constructivist learning approaches, which enhance learning by connecting material to relevant and relatable experiences for learners [42, 43]. One such example could be for instructors to present to their learners historical events as interconnected personal stories aiming to foster empathy and a deeper understanding of the events' significance.

Moreover, with the help of AI instructors can craft rich, multidimensional narratives that embed abstract concepts within meaningful, relatable contexts. AI automates much of the content production process, allowing educators to focus on weaving these elements into immersive stories that emotionally and intellectually resonate with learners. For instance, instructors can use AI-generated text summaries, interactive simulations, or dynamic videos as foundational components to build narratives that make complex topics, such as quantum mechanics or economic theories, more relatable, engaging, and memorable. AI also scaffolds instructors in becoming multimedia storytellers, enabling them to exploit and combine diverse formats, like text, visuals, audio, and interactivity into cohesive, engaging lessons. Thus, AI supports multimodal storytelling, which engages learners through multiple sensory channels and enhances accessibility across diverse student populations [44].

Similarly to directors crafting films that evoke emotions and leave lasting impressions, instructors must design learning journeys that challenge, motivate, and actively engage their students. Although, storytelling as an approach, was once the domain of highly skilled educators, now is rendered as more accessible due to AI facilitating the creation of narratives that enrich and transform the learning experience [41]. This combination of storytelling and technology lets teachers take on the roles of directors, writers, and producers, who create engaging and impactful learning experiences that resonate with learners on both intellectual and emotional levels.

3.4 Directing at a Meta-Level

AI will introduce a new dimension to instructional design, where instructors no longer solely create static lesson plans but instead design meta-descriptions of learning experiences. These meta-descriptions are abstract, high-level blueprints that outline the objectives, narratives, and structures of the learning experience. AI and learners then, will adjust and personalize these frameworks, tailoring the experience to individual needs and preferences. This is a shift toward a more abstract and strategic level of instructional design, in which instructors focus on creation of the blueprint upon which adaptive and personalized experiences are built.

In this meta-level role, the instructors become architects of adaptable learning frameworks that define the guiding principles, pathways, and key components of the educational journey. For example, instead of creating a single fixed lesson solely based on climate change, an instructor might author a meta-description that includes potential narratives (scientific, ethical, cultural), adaptable modules (interactive simulations, case studies, policy debates), and branching scenarios depending on learner input or AI-driven analytics.

Such a process puts AI in charge of doing the execution, pacing, difficulty, and content presenting, while learners actively shape their journey within the provided framework. The instructor sets the vision and offers a blueprint for exploration while keeping the experience cohesive, aligned with learning objectives, and meaningful across multiple contexts.

At this meta-level, instructors collaborate with AI as an active partner and not just as a tool. Instructors set the overall vision and structure, while AI manages the light adjustments and iterations based on real-time data. In turn, learners, participate as co-creators, navigating their paths within the designed instructor's framework. This collaborative dynamic allows for:

- Personalized learning experiences that reflect individual needs.
- Active involvement in the learning experience that leads to increased learner agency.
- Dynamic feedback loops, with AI continuously refining content delivery based on learner's progress and preferences.

This meta-level design challenges the educators to think beyond the construction of traditional lesson plans and instead to focus on how learning experiences may change dynamically with AI and learner collaboration.

3.5 *Directing Non-linear, Multimodal Experiences*

With the aid of AI instructors can create non-linear and multimodal learning experiences that engage learners through their different media formats, such as interactive videos, simulations, gamified assessments, and more. These experiences support a varied of learning styles and preferences providing learners with multiple paths and methods of engaging since they can to interact with the educational content in accordance to their unique interests and needs. Creating such complex and non-linear material requires a set of skill similar to that of a movie director managing multi-threaded storyline. Instructors must ensure that all the component—visual, audio, or interactive—work cohesively to advance the learning objectives. This orchestration requires a creative vision along with a deep understanding of both pedagogy and technology, blending them seamlessly to create immersive and effective learning environments.

Transformation through AI storytelling or AI-enhanced storytelling is impressive, as shown by [45], which explored how AI can assist in interactive and non-linear storytelling in cases such us role-playing games (e.g., Dungeons & Dragons). In these types of scenarios, AI can act as a dynamic storyteller, generating a variety of characters, adapting narratives in real-time, and creating vivid illustrations in order to enhance engagement and creativity. These findings highlight how AI advances creativity and decision-making, providing a framework for revolutionary applications in educational contexts. For example, a history instructor could design an interactive story where students could act as historical figures, making decisions that affect specific events and that way experience the consequences of their choices. On another example, a biology course could provide simulations that allow students to experiment in various ecological interventions and observe how ecosystems would be responding according to their actions. These applications let learners to actively adjust their learning experiences, according to their personal understanding of cause-and-effect relationships and the real-world complexities.

4 Closing Remarks

While AI handles many of the routine tasks, instructors involve themselves in activities that require a higher order such as strategic planning, ethical decision-making, and curating innovative learning paths. The fundamental educational shift happens when the instructor evolves into a director of AI-driven learning experiences. Just like director are responsible for the overall creative and strategic vision of a film so is the instructor regarding the overall narrative the learning journey is intended to follow. Thus instructors have the following responsibilities:

- Directing with Vision. Educational goals are clearly stated, with all content and all activities aligned to this vision.
- Exploring aggressively the design space: breadth-first design explorations, concurrently producing multiple instructional methods.
- Becoming charismatic storytellers: Conception of authentic, immersive stories that engage learners emotionally and intellectually.
- Directing at a Meta level: Creating adaptable frameworks that for personalized learning paths.
- Orchestrating Complexity: Designing non-linear, multimodal experiences catering to diverse learner needs.

The barriers that often stifled creativity-logistical complications, human cognitive biases, and time constraints are now a lesser concern in the age of AI. The educational vision, once impractical in the real-world of teaching, returns as a central tenet, allowing instructors to preserve the empathy, creativity, and ethical responsibility that define an impactful teaching. Yet this transition also means that teachers will need to acquire new skills for exploring different possibilities, asking the right questions to the students, and aligning choices with educational goals. What was once a sequential block-by-block design process is now a dynamic practice of rapid prototyping for flexible and adaptive learning scenarios.

To support this transition, teacher preparation programs should integrate AI literacy and instructional design strategies that emphasize the curation, orchestration, and direction of AI-generated content. Specifically, professional development should include: (a) scenario-based training to help educators explore breadth-first instructional design using AI tools; (b) reflective activities for aligning AI-generated content with pedagogical goals and ethical standards; and (c) practice in prompt engineering and adaptive decision-making based on real-time analytics. Moreover, institutional support mechanisms—such as mentoring systems, design sprints, and access to curated AI tools—are necessary to help instructors adopt the director role. These align with existing frameworks (e.g., UNESCO AI Competency Framework, EThICAL AI Readiness) which stress active, iterative engagement with AI and the need for a mindset shift toward strategic instructional design.

Looking ahead, the future for educators is striking the right balance between AI-driven automation and everything aspect of teaching that based on human aspects. Instructors must actively shape how these technologies will be used thus keeping control and ethical responsibility, and not simply adopt AI tools. Learning remains an inherently complex and demanding process and requires active cognitive engagement, sustained effort, and emotional resilience from the side of the learner. While AI brings many new efficiencies and advantages in instructional design, it does not change the fundamental nature of learning. True learning still depends on the learner's ability to fight with challenges, connect new knowledge to prior understanding, and persist mastery what has been learned. As we take advantage of the important opportunities provided by AI, we need to be aware that in reality learning will always require effort, self-regulation and intrinsic motivation from the learner.

References

1. Aslan, S., Reigeluth, C.M.: A trip to the past and future of educational computing: understanding its evolution. Contemp. Educ. Technol. **2**(1), 1–17 (2011). https://doi.org/10.30935/cedtech/6040
2. Reiser, R.A.: A history of instructional design and technology: part I: a history of instructional media. Educ. Technol. Res. Dev. **49**(1), 57–67 (2001). https://doi.org/10.1007/BF02504928
3. Saettler, P.: The Evolution of American Educational Technology. IAP, Charlotte (2004)
4. Selwyn, N.: Digital degrowth: toward radically sustainable education technology. Learn. Media Technol. **49**(2), 186–199 (2024). https://doi.org/10.1080/17439884.2022.2159978
5. Romiszowski, A.J.: How's the e-learning baby? Factors leading to success or failure of an educational technology innovation. Educ. Technol. **44**(1), 5–27 (2004)
6. Du, H., et al.: Harnessing large language models to auto-evaluate the student project reports. Comput. Educ. Artif. Intell. **7**, 100268 (2024). https://doi.org/10.1016/j.caeai.2024.100268
7. Fokides, E., Peristeraki, E.: Comparing ChatGPT's correction and feedback comments with that of educators in the context of primary students' short essays written in English and Greek. Educ. Inf. Technol. (2024). https://doi.org/10.1007/s10639-024-12912-8
8. Guo, K., Wang, D.: To resist it or to embrace it? Examining ChatGPT's potential to support teacher feedback in EFL writing. Educ. Inf. Technol. **29**(7), 8435–8463 (2024). https://doi.org/10.1007/s10639-023-12146-0
9. Chen, X., Wu, D.: Automatic generation of multimedia teaching materials based on generative AI: taking tang poetry as an example. IEEE Trans. Learn. Technol. (2024). https://doi.org/10.1109/TLT.2024.3378279
10. Moundridou, M., Matzakos, N., Doukakis, S.: Generative AI tools as educators' assistants: designing and implementing inquiry-based lesson plans. Comput. Educ. Artif. Intell. **7**, 100277 (2024). https://doi.org/10.1016/j.caeai.2024.100277

11. Sun, L., Zhou, L.: Does generative artificial intelligence improve the academic achievement of college students? A meta-analysis. J. Educ. Comput. Res. **62**(7), 1896–1933 (2024). https://doi.org/10.1177/07356331241277937
12. Yilmaz, R., Yilmaz, F.G.K.: The effect of generative artificial intelligence (AI)-based tool use on students' computational thinking skills, programming self-efficacy, and motivation. Comput. Educ. Artif. Intell. **4**, 100147 (2023). https://doi.org/10.1016/j.caeai.2023.100147
13. Mikeladze, T., Meijer, P.C., Verhoeff, R.P.: A comprehensive exploration of artificial intelligence competence frameworks for educators: a critical review. Eur. J. Educ. (2024). https://doi.org/10.1111/ejed.12663
14. Miao, F., Cukurova, M.: AI Competency Framework for Teachers. UNESCO, Paris (2024). https://doi.org/10.54675/ZJTE2084
15. Celik, I.: Towards intelligent-TPACK: an empirical study on teachers' professional knowledge to ethically integrate artificial intelligence (AI)-based tools into education. Comput. Hum. Behav. **138**, 107468 (2023). https://doi.org/10.1016/j.chb.2022.107468
16. Filo, Y., Rabin, E., Mor, Y.: An artificial intelligence competency framework for teachers and students: co-created with teachers. Eur. J. Open Distance E-Learn. **26**(1) (2024). https://doi.org/10.2478/eurodl-2024-0012
17. Luckin, R., Cukurova, M., Kent, C., Du Boulay, B.: Empowering educators to be AI-ready. Comput. Educ. Artif. Intell. **3**, 100076 (2022). https://doi.org/10.1016/j.caeai.2022.100076
18. Ng, D.T.K., Leung, J.K.L., Su, J., Ng, R.C.W., Chu, S.K.W.: Teachers' AI digital competencies and twenty-first century skills in the post-pandemic world. Educ. Technol. Res. Dev. **71**(1), 137–161 (2023). https://doi.org/10.1007/s11423-023-10203-6
19. Palaigeorgiou, G., Christodoulou, P.: AI and the art of instructional design: exploring 5 layers of expertise. LearnWorlds. (2024). https://www.learnworlds.com/ai-instructional-design-ebook/
20. Choi, G.W., Kim, S.H., Lee, D., Moon, J.: Utilizing generative AI for instructional design: exploring strengths, weaknesses, opportunities, and threats. TechTrends. (2024). https://doi.org/10.1007/s11528-024-00967-w
21. Chng, L.K.: How AI makes its mark on instructional design. Asian J. Distance Educ. **18**(2), 32–41 (2023). http://asianjde.com/ojs/index.php/AsianJDE/article/view/740
22. Parsons, B., Curry, J.H.: Can ChatGPT pass graduate-level instructional design assignments? Potential implications of artificial intelligence in education and a call to action. TechTrends. **68**, 67–78 (2024). https://doi.org/10.1007/s11528-023-00912-3
23. Ruiz-Rojas, L.I., Acosta-Vargas, P., De-Moreta-Llovet, J., Gonzalez-Rodriguez, M.: Empowering education with generative artificial intelligence tools: approach with an instructional design matrix. Sustainability. **15**(15), 11524 (2023). https://doi.org/10.3390/su151511524
24. Alqahtani, T., Badreldin, H.A., Alrashed, M., Alshaya, A.I., Alghamdi, S.S., Bin Saleh, K., Albekairy, A.M.: The emergent role of artificial intelligence, natural language processing, and large language models in higher education and research. Res. Soc. Adm. Pharm. **19**(8), 1236–1242 (2023). https://doi.org/10.1016/j.sapharm.2023.05.016
25. Dickey, E., Bejarano, A.: GAIDE: A Framework for Using Generative AI to Assist in Course Content Development. arXiv preprint arXiv:2308.12276 (2024)
26. McNeill, L.: Automation or innovation? A generative AI and instructional design snapshot. IICE Off. Conf. Proc. (2024). https://doi.org/10.22492/issn.2189-1036.2024.17
27. Bolick, A.D., Da Silva, R.L.: Exploring artificial intelligence tools and their potential impact to instructional design workflows and organizational systems. TechTrends. **68**(1), 91–100 (2024). https://doi.org/10.1007/s11528-023-00894-2
28. Luo, T., Muljana, P.S., Ren, X., Young, D.: Exploring instructional designers' utilization and perspectives on generative AI tools: a mixed methods study. Educ. Technol. Res. Dev. (2024). https://doi.org/10.1007/s11423-024-10437-y
29. Pesovski, I., Santos, R., Henriques, R., Trajkovik, V.: Generative AI for customizable learning experiences. Sustainability. **16**, 3034 (2024). https://doi.org/10.3390/su16073034
30. Cohn, C., Snyder, C., Fonteles, J.H., Montenegro, J., Biswas, G.: A multimodal approach to support teacher, researcher and AI collaboration in STEM+C learning environments. Br. J. Educ. Technol. (2024). https://doi.org/10.1111/bjet.13518

31. Dai, C.P., Ke, F., Pan, Y., Moon, J., Liu, Z.: Effects of artificial intelligence-powered virtual agents on learning outcomes in computer-based simulations: a meta-analysis. Educ. Psychol. Rev. **36**(1), 31 (2024). https://doi.org/10.1007/s10648-024-09855-4
32. Verhelst, E., Janssens, R., Demeester, T., Belpaeme, T.: Adaptive second language tutoring using generative AI and a social robot. In: Companion of the 2024 ACM/IEEE International Conference on Human-Robot Interaction, pp. 1080–1084. ACM, New York (2024). https://doi.org/10.1145/3610978.3640559
33. Clow, D.: An overview of learning analytics. Teach. High. Educ. **18**(6), 683–695 (2013). https://doi.org/10.1080/13562517.2013.827653
34. Viberg, O., Hatakka, M., Bälter, O., Mavroudi, A.: The current landscape of learning analytics in higher education. Comput. Hum. Behav. **89**, 98–110 (2018). https://doi.org/10.1016/j.chb.2018.07.027
35. Yan, L., Martinez-Maldonado, R., Gasevic, D.: Generative artificial intelligence in learning analytics: Contextualising opportunities and challenges through the learning analytics cycle. In: Proceedings of the 14th Learning Analytics and Knowledge Conference, pp. 101–111. ACM, New York (2024). https://doi.org/10.1145/3636555.3636856
36. Cukurova, M.: The interplay of learning, analytics and artificial intelligence in education: a vision for hybrid intelligence. Br. J. Educ. Technol. (2024). https://doi.org/10.1111/bjet.13514
37. Javed, R., Nasir, O., Borit, M., Vanhée, L., Zea, E., Gupta, S., Vinuesa, R., Qadir, J.: Get out of the BAG! Silos in AI ethics education: unsupervised topic modeling analysis of global AI curricula. J. Artif. Intell. Res. **73**, 933–965 (2022). https://doi.org/10.1613/jair.1.13550
38. Stilgoe, J., Owen, R., Macnaghten, P.: Developing a framework for responsible innovation. Res. Policy. **42**(9), 1568–1580 (2013). https://doi.org/10.1016/j.respol.2013.05.008
39. Stahl, B.C., Wright, D.: Ethics and privacy in AI and big data: implementing responsible research and innovation. IEEE Secur. Privacy. **16**(3), 26–33 (2018). https://doi.org/10.1109/MSP.2018.2701164
40. Andrews, D.H., Hull, T.D., Donahue, J.A.: Storytelling as an instructional method: Definitions and research questions. Interdiscip. J. Probl. Based Learn. **3**(2), 1063 (2009). https://doi.org/10.7771/1541-5015.1063
41. Smeda, N., Dakich, E., Sharda, N.: The effectiveness of digital storytelling in the classrooms: a comprehensive study. Smart Learn. Environ. **1**, 1–21 (2014). https://doi.org/10.1186/s40561-014-0006-3
42. Wu, J., Chen, D.T.V.: A systematic review of educational digital storytelling. Comput. Educ. **147**, 103786 (2020). https://doi.org/10.1016/j.compedu.2019.103786
43. Jonassen, D.H., Hernandez-Serrano, J.: Case-based reasoning and instructional design: using stories to support problem solving. Educ. Technol. Res. Dev. **50**(2), 65–77 (2002). https://doi.org/10.1007/bf02504994
44. Arif, S., Arif, T., Haroon, M.S., Khan, A.J., Raza, A.A., Athar, A.: The art of storytelling: multi-agent generative AI for dynamic multimodal narratives. arXiv preprint arXiv:2409.11261 (2024). https://doi.org/10.48550/arXiv.2409.11261
45. Santiago III, J.M., Parayno, R.L., Deja, J.A., Samson, B.P.V.: Rolling the dice: Imagining generative AI as a dungeons & dragons storytelling companion. arXiv preprint arXir: 2409.11261 (2023). https://doi.org/10.48550/arXiv.2409.11261

TSM-SGP: Taxonomy of Support Methods in Serious Games for Programming

Pavlos Toukiloglou and Stelios Xinogalos

Abstract Serious games (SGs), designed for purposes beyond entertainment, have gained popularity in engaging students in programming education. One crucial aspect of SGs is a support system, which is pivotal in achieving their educational objectives. However, the lack of consensus in the literature regarding the terminology and classification of support systems results in inconsistencies and varying interpretations among researchers. This ongoing research presents the Taxonomy of Support Methods for Serious Games for Programming (TSM-SGP) to address this gap. This taxonomy offers a systematic framework for identifying and categorizing support systems specifically within the programming domain. It offers clarity, enabling researchers, educators, and game designers to gain a deeper understanding of their roles and effectiveness. Additionally, the taxonomy reveals insightful relationships between support methods and their properties, contributing to the development of effective SGs.

Keywords Taxonomy · Serious games · Support methods · Programming

1 Introduction

Serious games (SGs), often referred to as educational games, are designed with objectives that extend beyond entertainment. The concept of SGs was first mentioned by Elder [1] and defined as games that have an "*explicit and carefully thought-out educational purpose and are not intended to be played primarily for amusement.*" The term "serious" encompasses video games utilized across various sectors, including defence, education, scientific exploration, healthcare, and more [2]. These games

P. Toukiloglou (✉) · S. Xinogalos
Department of Applied Informatics, School of Information Sciences, University of Macedonia, Thessaloniki, Greece
e-mail: toukiloglou@uom.edu.gr

S. Xinogalos
e-mail: stelios@uom.edu.gr

I. Kazanidis and A. Tsinakos (eds.), *Research on E-Learning and ICT in Education*,
https://doi.org/10.1007/978-3-031-99961-1_9

have the potential to promote a deep understanding of specific topics and be an effective learning tool [3]. SGs are gaining popularity as a tool for delivering educational content to students [4]. Their ability to enhance engagement and provide immersive learning experiences makes them a compelling choice in educational settings.

Computational thinking (CT), on the other hand, has become a fundamental skill of analytical thinking for everyone in the twenty-first century [5]. It encompasses the utilization of fundamental computer science principles, including abstraction, debugging, and problem-solving. Programming promotes CT since it serves as a scaffold for its development [6]. Regardless, programming is a complex task, posing significant challenges in the teaching and learning process. To address this problem and enhance students' chances of success, SGs emerged as a promising option. The effectiveness of SGs as learning tools relies heavily on the presence of robust support systems and methods that aid learners in achieving their educational objectives. These systems are essential for promoting knowledge and skill acquisition by providing learners with necessary scaffolding throughout the educational process. They include various features and mechanisms designed to offer guidance, feedback, and reinforcement, ensuring that learners receive appropriate support customized to their individual needs and progress. Without well-designed support, SGs may fail to deliver on their educational promises, hindering both cognitive and skill development. Consequently, a comprehensive understanding of the diverse types of support systems and methods is critical for maximizing their educational potential.

This study proposes a taxonomy to methodically identify and present the properties of support methods in SGs for programming. A taxonomy is a systematic framework used for classifying, organizing, and categorizing elements within a specific domain based on their shared characteristics and relationships. It provides a structured approach to managing complex information, promoting clarity, consistency, and effective communication within a field. In the context of educational technologies and SGs, a well-defined taxonomy offers numerous benefits. It enables researchers and practitioners to systematically identify and classify various components and features, thus enhancing the precision and clarity of academic discourse. Additionally, taxonomies support standardized terminology development, promoting common understanding and reducing uncertainties across different studies and applications. This standardization is essential for comparative studies and meta-analyses, as it allows for the aggregation and synthesis of findings from diverse sources. Furthermore, taxonomies guide the design and implementation of educational interventions by providing a comprehensive overview of possible options and their interrelations. This taxonomy aims to explore the characteristics, functionalities, and roles that support methods play in enhancing the learning experience in SGs for programming. The proposed new taxonomical scheme, called Taxonomy of Support Methods for Serious Games for Programming (TSM-SGP) establishes standardized terminology for describing support methods and contributes to the field through the following key advancements:

– Identification: The TSM-SGP systematically identifies and categorizes the various types of support methods within the context of SGs for programming.

- Terminological Precision: TSM-SGP promotes clear and precise communication within the scientific literature.
- Addressing Literature Gaps: The TSM-SGP addresses a gap in the existing literature by providing a structured framework that was previously lacking.
- Clarity and Structure: The taxonomy promotes a common understanding among serious game designers, developers, academics, and students regarding the available support options within these games. This shared clarity is important for the collaborative development and implementation of effective support methods.

The study is structured as follows: Sect. 2 provides an overview of existing literature on support methods in SGs. Section 3 outlines the methodology used for developing the taxonomy. Section 4 provides a detailed classification of diverse support methods, highlighting their distinct characteristics and applications. In Sect. 5, the results are analysed, and use cases of the taxonomy are illustrated, offering insights into future research directions and opportunities for advancing SGs support methods. Lastly, Sects. 6 and 7 present a discussion and conclusion about the proposed taxonomy.

2 Related Work

The literature on taxonomies of support systems in SGs is limited in scope, with no consensus on their characterization and categorization. While support systems are acknowledged as key components in other frameworks related to educational technology and interactive learning environments, a dedicated taxonomy is lacking. This section reviews existing taxonomies and their implications for support systems in SGs, laying the foundation for the proposed framework.

Bedwell et al. [7] introduced a taxonomy categorizing SG attributes to enable comparisons between learning outcomes in games and traditional methods. Although lacking a specific category for support systems, the taxonomy included "assessment" as a key attribute, describing feedback provided to guide player actions. The study concluded that feedback is a critical component in achieving educational objectives in SGs. De Lope and Medina [8] proposed a taxonomy with 16 criteria covering various game features, including "assessment" and "adaptation." Assessment refers to explicit or implicit evaluation methods to gauge player learning outcomes, while adaptation describes strategies to customize game elements to player abilities. These aspects partially relate to support systems but do not address them comprehensively. Laine and Lindberg [9] conducted a systematic review of educational game design, presenting a taxonomy linking game motivators to design principles. Feedback emerged as a crucial motivator, with instructions, tutorials, and prompts enhancing engagement. Prompt and goal-oriented feedback was particularly effective, highlighting its role in supporting players during gameplay challenges.

Rato and Prada [10] proposed a taxonomy of social roles for agents in games, emphasizing the Tutor/Learner NPC (non-player character). These NPCs support players by providing guidance, tutorials, and strategies, delivered proactively or upon request. The study highlighted how interactions with such agents improve player understanding of game mechanics and overall experience. Hamrouni and Bendella [11] developed a taxonomy categorizing emotions in SGs into beneficial and detrimental types. Activating emotions like frustration drive motivation while deactivating emotions like boredom hinder engagement. The findings suggest that emotional dynamics significantly impact the effectiveness of SGs, underlining the importance of integrating emotional support mechanisms. While existing taxonomies address related aspects such as feedback, adaptation, and social interactions, none provide a comprehensive framework dedicated to support systems. This gap underscores the need for a unified taxonomy systematically identifying and categorizing support mechanisms within SGs for programming. The proposed taxonomy builds on these insights, integrating relevant concepts to provide a robust and practical framework for support systems in SGs.

3 Methodology

The development of TSM-SGP was based on the methodological framework proposed by Nickerson, Varshney, and Muntermann [12]. Their approach combines conceptual and empirical cycles that provide a clear foundation for defining and refining the categories and elements within TSM-SGP. The method defines taxonomy as a set of N dimensions Di ($i = 1,...,n$) where each dimension consists of Kj (Kj ≥ 2) characteristics Cij. In this context, a dimension refers to a distinct feature or perspective used to classify objects, ensuring consistent grouping based on shared characteristics. Each characteristic is mutually exclusive, ensuring that each support description has exactly one Cij for each Di. This restriction ensures that when using the taxonomy to describe a support system, no support description can have two different characteristics in the same dimension, therefore creating a complete and non-overlapping classification system. The method involves iterative steps, with ending conditions being evaluated after each iteration for ensuring the taxonomy's comprehensiveness and refinement. The seven steps of the taxonomy development method are presented in Fig. 1.

Step 1: Determine the meta-characteristic. This step is critical as it determines the central concept that impacts every succeeding classification decision. It identifies the most comprehensive characteristic that will serve as the basis for choosing relevant attributes in the taxonomy. For TSM-SGP, this meta-characteristic is "support methods in serious games for programming."

Step 2: Determine Ending Conditions. According to the methodology, there are two types of ending conditions to determine when the taxonomy development process should end. Objective conditions include criteria such as the absence of newly introduced dimensions in the final iteration, while subjective conditions ensure the taxonomy is concise, robust, comprehensive, extendible, and explanatory. Both

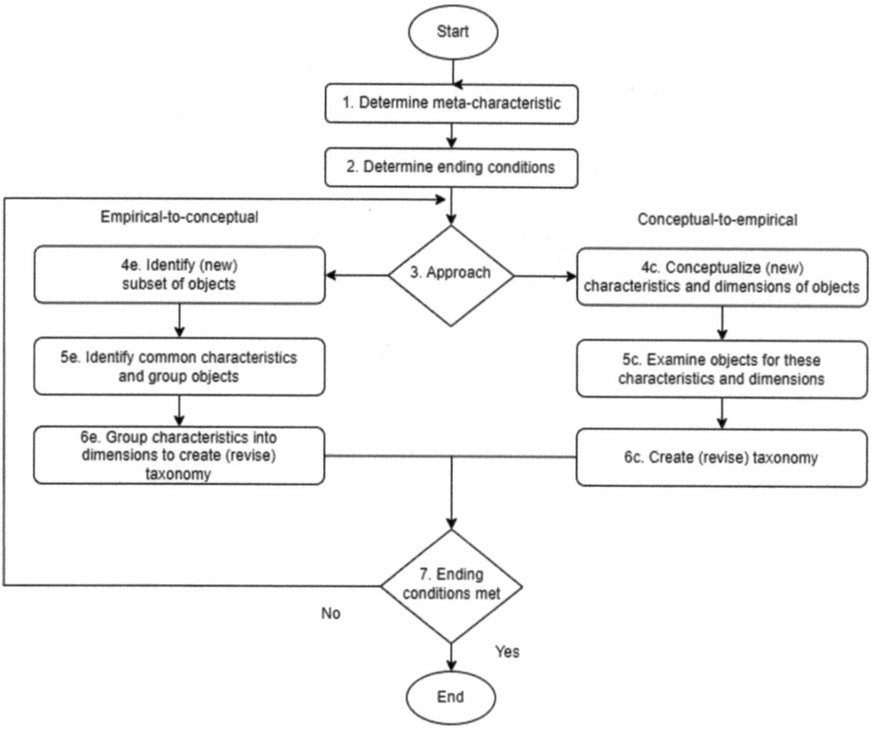

Fig. 1 The taxonomy development methodology

objective and subjective conditions are presented in Table 1. For TSM-SGP, all conditions for both types were satisfied.

Step 3: Choose Approach. Nickerson et al. [12] outline two primary approaches. The empirical-to-conceptual approach begins with data (e.g., existing examples or cases) and derives dimensions and characteristics from observed patterns. The conceptual-to-empirical approach starts with theoretical or conceptual insights and subsequently validates these through empirical data. The choice depends on the availability of data and the researcher's knowledge of the domain. For TSM-SGP, both approaches were employed iteratively. Initially, the empirical-to-conceptual approach was adopted, utilizing the domain knowledge and available data from existing SGs for programming to identify preliminary dimensions and characteristics. Once no further new attributes could be extracted, the process shifted to the conceptual-to-empirical approach, where the emerging classification framework was refined against theoretical constructs and literature. This alternation continued, reverting to empirical examination when necessary until all ending conditions were satisfied.

Table 1 Objective and subjective ending conditions

Objective ending conditions	Subjective ending conditions
All objects or a representative sample of objects have been examined	Concise (does the number of dimensions allow the taxonomy to be meaningful without being unwieldy or overwhelming?)
No object was merged with a similar object or split into multiple objects in the last iteration	Robust (do the dimensions and characteristics provide for differentiation among objects sufficient to be of interest?)
At least one object is classified under every characteristic of every dimension	Comprehensive (can all objects or a (random) sample of objects within the domain of interest be classified?)
No new dimensions or characteristics were added in the last iteration	Extendible (can a new dimension or a new characteristic of an existing dimension be easily added?)
No dimensions or characteristics were merged or split in the last iteration	Explanatory (what do the dimensions and characteristics explain about an object?)
Every dimension is unique and not repeated	
Every characteristic is unique within its dimension	
Each cell (combination of characteristics) is unique and is not repeated	

Step 4e: Identify a Subset of Objects (Empirical). During this phase, existing support methods were selected for analysis. To ensure the validity and representativeness of the taxonomy, the selection of SGs complied with the following criteria:

- Academic foundation: Games must be referenced in peer-reviewed academic literature or official documentation to ensure alignment with scholarly standards.
- Programming focus: Games must explicitly target programming or computational thinking skills.
- Accessibility: Games must be either (a) publicly available for reproducibility or (b) described in sufficient detail.
- Support methods: Games must employ at least one distinct support method.

Step 5e: Identify Common Characteristics and Group Objects (Empirical). The selected support methods were examined and categorized based on observable similarities and differences. During this process, new characteristics were identified, and existing ones refined.

Step 6e: Search for New Characteristics and/or Dimensions (Empirical). The current set of characteristics and dimensions was assessed to ensure they sufficiently classified all support methods. If new traits or groupings emerged, the taxonomy was expanded or revised accordingly.

Step 4c: Conceptualize (New) Characteristics and Dimensions of Objects. During this phase, new characteristics and dimensions were proposed or refined based on domain expertise, prior research, and established theories.

Step 5c: Examine objects for these characteristics and dimensions. The taxonomy's elements were evaluated to determine whether they aligned with the proposed characteristics and dimensions.

Step 6c: Create (revise) taxonomy. Based on insights gained from the evaluation, the taxonomy was refined by incorporating any necessary modifications to the characteristics and dimensions.

Step 7: During this phase, the taxonomy was evaluated against the objective and subjective criteria defined in Step 2 to determine whether the development process was complete. If all ending conditions were met, the process concluded. However, if any of these conditions remained unmet, development returned to Step 3 to select an appropriate approach.

According to the framework proposed by Nickerson et al. [12], the TSM-SGP development methodology satisfies the following criteria:

- It employs multiple strategies to construct the taxonomy, integrating both inductive (empirical) and deductive (conceptual) processes.
- It minimizes the possibility of introducing arbitrary dimensions and characteristics into the taxonomy by requiring a systematic process for identifying and defining each element.
- It is designed to reach a defined endpoint within a limited number of iterations, therefore concluding the process in a reasonable timeframe.
- The approach provides a clear sequence of steps that is relatively simple to implement.
- It produces a practical and useful taxonomy.

Through these structured steps, TSM-SGP's development demonstrates a precise and flexible use of the adopted methodology. The process combines empirical observations with theoretical insights, ensuring that each characteristic is valid and systematically integrated. The iterative approach produces a balanced framework that minimizes arbitrary inclusions and provides a straightforward set of procedures to guide its construction. The result is a practical and valuable taxonomy that classifies support methods in SGs for programming using a method that is both comprehensive and adaptable, setting a foundation for future research and development in this domain.

4 Classification Criteria

The classification is structured into six primary dimensions (categories), each describing a distinctive characteristic. For each dimension, the specific sub-elements and functions are described in terms of their effectiveness, strengths, and weaknesses.

4.1 Integration

Integration refers to the method or environment in which support is provided within SGs. Support mechanisms are classified as internal or external, each offering distinct assistance methods. Internal support is embedded in the game's interface, allowing learners to receive help without leaving the environment. This maintains immersion and seamless interaction. Examples include in-game tutorials introducing mechanics and content, pop-up tips offering task-specific hints, and contextual help like tooltips or manuals accessed on demand. External support, accessed outside the game, complements learning with resources such as websites, detailed tutorials, forums, and physical materials like guides or flashcards. Floating windows, part of the game but separate from its world, display extra information without disrupting gameplay, enabling access to external sources while maintaining immersion. Integration design offers benefits and challenges. Internal integration enhances immersion by keeping learners in the game but can be hard to update without altering the game environment. Solutions like importing tutorial data independently require extra programming and work within predefined resource limits. In contrast, external integration allows easy updates but may disrupt immersion by shifting learners between the game and external resources. Floating windows balance the two but can distract if poorly designed. Each approach's suitability depends on educational goals and game design, aiming to ensure effective learning support.

4.2 Media

The media criterion includes five formats: text, images, video, audio, and animation. Webpages, combining these formats, often act as external support sources, hosted online or locally. Remote hosting requires constant uptime, increasing costs for long-term use and server maintenance for many users. If developers do not create the webpage, its content may change unpredictably, affecting the quality and reliability of the educational material.

Text is a universal and adaptable support method in SGs, using written language to present information. Early SGs relied on text due to technological limits. Text's flexibility allows customization through formatting and style, offering searchability, editability, and upgradability. It is also memory-efficient, enabling quick access to large data volumes. However, text lacks the immersive qualities associated with visual or auditory media. In educational contexts, students often prefer interactive or multimedia formats, as static text can detract from immersion. Text-heavy support may challenge learners with limited reading skills or language proficiency, particularly young students, requiring careful terminology and presentation. Conveying complex emotions or tones via text is difficult. Developers address this with enhancements like colour formatting, emojis, special characters, or animations (e.g., typewriter effect in AI for Oceans, [13]), compensating for text's limitations.

Images are widely used in SGs to deliver visual instructions and information compactly and clearly. They simplify complex concepts, enhance immersion, and break up text-heavy environments, making learning more engaging. Images help visualize abstract ideas, such as flowchart execution or data structures like binary trees and arrays, aiding comprehension, especially for visual learners. They also integrate educational content with the game's narrative, boosting motivation by aligning material with themes. Images provide contextual information beyond text, illustrating programming commands like directions, loops, and actions to connect terms with meanings. However, their use must be balanced to avoid cognitive overload or distraction. Poor-quality or irrelevant images can detract from learning, emphasizing the need for high-quality, relevant visuals. Accessibility is critical, as visually impaired learners may need alternatives like descriptions or assistive technologies. Additionally, high-resolution images may strain system performance on low-end hardware or mobile devices, potentially affecting gameplay.

Video support provides a multisensory experience by combining visual and auditory elements to present educational content in SGs. Videos demonstrate processes, illustrate real-time code execution, and add narrative depth to learning. They can play full-screen, pausing gameplay, or in a window alongside game elements. Players can pause, rewind, or stop videos to aid comprehension, with options to skip familiar content or adjust playback speed for personalized learning. These features make videos dynamic and engaging tools for content delivery. However, producing video content is time-intensive and costly, requiring significant resources for scripting, filming, and editing. Videos also demand substantial storage, and compression can affect quality. On low-end hardware, they increase memory and processing requirements, potentially impacting game performance.

Audio support is delivered through voice-overs, sound effects, and narration, enhancing immersion and emotional connection when visuals are insufficient. Sound effects provide feedback on correct answers or actions, enriching interactivity by signalling success or guiding tasks. Voice-overs and narration offer instructions, context, and storytelling, aiding comprehension and engagement. However, learners with hearing impairments or limited language proficiency may face challenges, necessitating alternatives like subtitles or sign language for accessibility. Producing high-quality audio requires costly equipment, skilled musicians, and professional voice actors, increasing project expenses and timelines. In shared environments, audio can disturb others unless individual headphones are provided. This can be addressed by offering robust volume controls for music, effects, and voice channels or ensuring players have headphones.

Animation is a dynamic visual medium, distinct from videos, as it uses the game's assets, like sprites and 3D models, seamlessly integrated into the environment. Unlike pre-recorded videos, animations are rendered in real-time, enabling interactive and responsive experiences. They provide on-demand instructions customized to player actions and effectively illustrate complex programme flows or command sequences. However, integrating animations with game elements adds complexity, requiring coordination with existing mechanics and assets. Proper planning and execution are essential to align animations with educational goals and enhance gameplay.

4.3 Technique

The technique element identifies the approaches SGs use to deliver learning content. While multiple techniques may be employed, one usually defines the core of the support system. Each technique has strengths and challenges, significantly impacting the educational experience.

Tips offer targeted suggestions or recommendations within SGs, typically as short messages or pop-ups. They provide support for navigation, task completion, or problem-solving, often offering increasing levels of assistance based on user requests. For instance, Dance Party [14] allows players to adjust the number of tips to suit their needs. However, overusing tips can disrupt gameplay, reducing challenge and engagement.

Manuals provide in-depth information on specific topics in SGs, often in digital format. Authored by experts, they ensure accurate, up-to-date content and offer a structured approach to complex subjects. However, manuals can be lengthy and dense, overwhelming learners who prefer interactive or visual support. Their static nature may also be less engaging than dynamic methods like tutorials or interactive guides.

Work examples demonstrate tasks or concepts, helping learners observe and understand strategies to address challenges in the game. Players can analyse these examples and apply the knowledge in similar scenarios. In programming, they illustrate problem-solving processes, aiding comprehension of complex concepts and algorithms [15]. Worked examples improve understanding and outcomes by guiding students through steps and showcasing coding best practices. Although relatively new in SGs for programming, tools like NanoDoc [16] have shown promising results in enhancing learning efficiency.

Chat allows players to communicate with NPCs or other players for assistance. NPC chat can be scripted, using dialogue trees for structured support [17], or dynamic, employing generative AI chatbots for personalized and immersive interactions [18]. Player-to-player chat typically involves brief, focused messages addressing specific topics, common in multiplayer games. However, human interactions can introduce distractions that detract from educational objectives. Proper design, moderation, and guidance are essential to ensure chats remain aligned with educational goals while promoting valuable insights and collaboration.

Narration uses audio to deliver learning content in SGs, implemented through NPC dialogue or a dedicated narrator. Seamlessly integrated into gameplay, it ensures immersion while providing direction, and context, highlighting key events, or reinforcing learning points, depending on the game's objectives.

4.4 Source

The source of support refers to the origin of educational material in the game, as defined by the taxonomy. The three main sources are expert, artificial intelligence, and collaboration, which can be used independently or combined in serious games. Human experts in programming, education, and academia are the most common sources of support material. They create educational content and work closely with designers to determine its presentation, making them essential to development. Experts provide broad information filtered to suit the game's target age group and difficulty. However, they cannot predict every player's question or issue. Employing experts is costly, and breaking content into on-demand sections requires significant time.

Advances in computer science are transforming the creation of support content in SGs. Technologies like natural language processing and machine learning now enable automation, with AI APIs generating content without direct human involvement. Models like GPT-4.5 [19] and LaMDA [20] produce human-like responses, enhancing interactivity and responsiveness in SGs. However, their focus on text-based interactions limits support forms like visual or auditory aids. AI-generated content, while scalable and efficient, risks inaccuracies without human fact-checking. As these technologies evolve, they will increasingly complement human expertise, creating dynamic, adaptive support methods.

Collaborative support arises from player interactions, common in multiplayer games with team-based assistance. Players may observe teammates' actions, provide guidance, or work together to solve problems on a shared screen [21]. Communication is managed by the game's infrastructure, with actions and support levels defined by the designer, educational objectives, and game type. Collaborative support enhances learning by promoting teamwork, communication, and problem-solving skills, leveraging the collective knowledge of players for a dynamic learning environment. However, its effectiveness depends on the game's design, ensuring alignment with educational goals and balancing guidance with independent problem-solving.

4.5 Individualization

Individualization classifies support as adaptive or non-adaptive, based on the player's knowledge. Adaptive support provides personalized assistance customized to the learner's progress and performance, dynamically adjusting content using approaches like player knowledge models or neural networks [22]. This ensures relevant and effective learning. Non-adaptive support, by contrast, is generic and standardized, not accounting for individual knowledge levels. While less effective, non-adaptive support is faster and easier to implement, making it suitable when personalized options are unfeasible due to resource constraints.

4.6 Engagement

Engagement categorizes player interaction with support as interactive or non-interactive. Interactive systems involve active player engagement, requiring actions or decisions during content delivery. While more immersive and engaging, they are harder to implement due to the need for effective tracking and responses. Non-interactive systems present content first, and then allow players to act, offering structured and consistent learning experiences. Easier to design, they are more common and safer for managing unpredictable player actions, ensuring all players receive the same information before engaging with the game.

5 Taxonomy Syntax

A summary of the key factors outlined in the proposed taxonomy is depicted in Fig. 2, providing a comprehensive overview of support methods. Following the established syntax within the taxonomy, developers and researchers can precisely identify support methods by incorporating the relevant techniques, media terms, and any necessary prefixes. The flexibility of the support design allows each term of the taxonomy to be utilized without strict syntax or order of appearance, as long as it thoroughly describes the implementation. In instances where multiple support systems and methods are present within a game, each description can be separated by a comma for clarity.

TSM-SGP demonstrates value when applied to existing studies, enhancing precision and clarity in describing support methods. For example, in the case of DungeonClass [16] the taxonomy allows for a more accurate description of the support method as "non-interactive animated worked examples" instead of the broader term "worked examples." This level of detail provides a clearer picture of the support system's nature and implementation. Similarly, the study on adaptive support using Neverwinter Nights [17] benefits from the taxonomy's precision. The support types, initially described as "non-adaptive text" and "adaptive text," can be more accurately characterized as "NPC text dialogue" and "adaptive NPC text dialogue," respectively. This refinement in terminology aligns more closely with the actual implementation of the support system and offers readers a more accurate understanding of the support mechanism. In the case of Minerva [23] by refining the initial description of "text, images, and video" to "instructional text, image hints, and worked examples video," TSM-SGP allows for a better understanding of each support element's specific purpose and implementation.

These examples highlight how TSM-SGP can enhance the clarity and precision of descriptions in scientific papers, potentially leading to improved communication and understanding within the research community. Additionally, the taxonomy's ability to capture subtle differences in support implementation can assist in more accurate comparisons between studies and support more robust meta-analyses in the field.

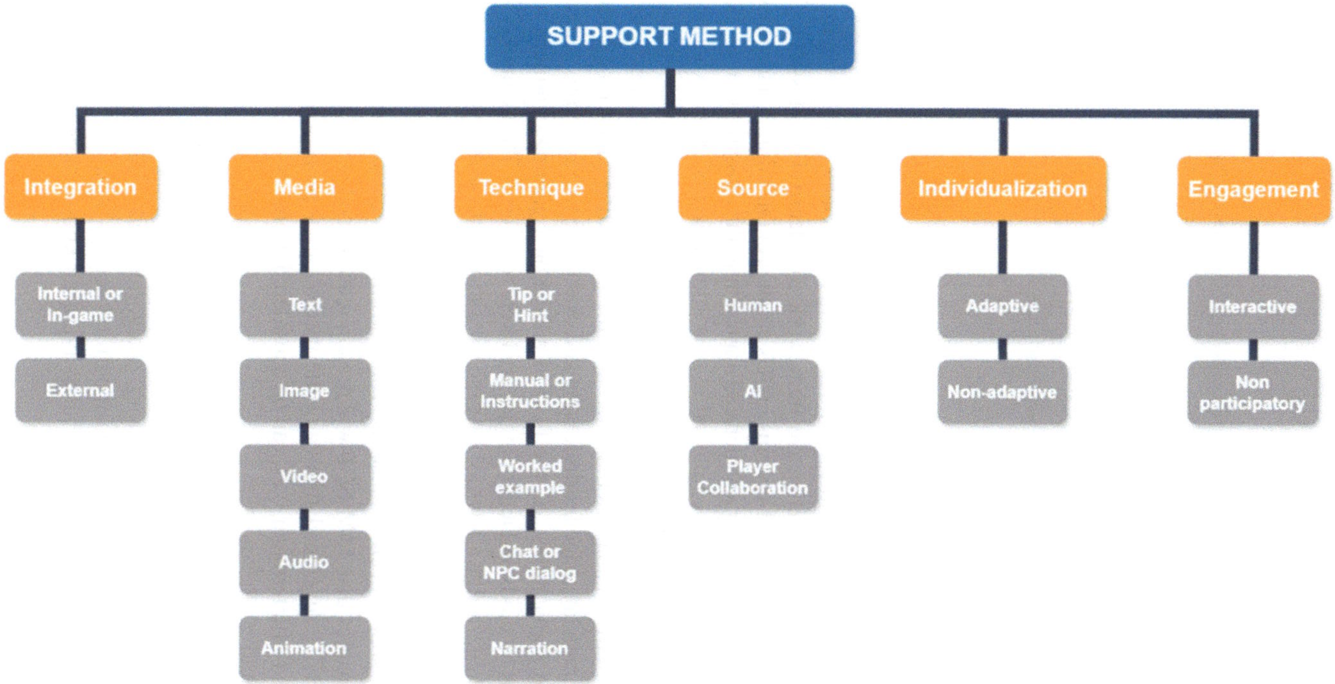

Fig. 2 The taxonomy factors

Some SGs employ multiple support methods in combination to assist users with specific game components, such as explaining new concepts or clarifying interface mechanics. A notable example is Dance Party [14], a popular game on the Hour of Code platform. As illustrated in Fig. 3, while individual support elements appear simple and mostly text-based, their collective implementation demonstrates a layered support system. Applying the TSM-SGP syntax, these methods can be categorized as follows: Text hints (Fig. 3a), Interactive audio and text instructions (Fig. 3b), External video instructions (Fig. 3c).

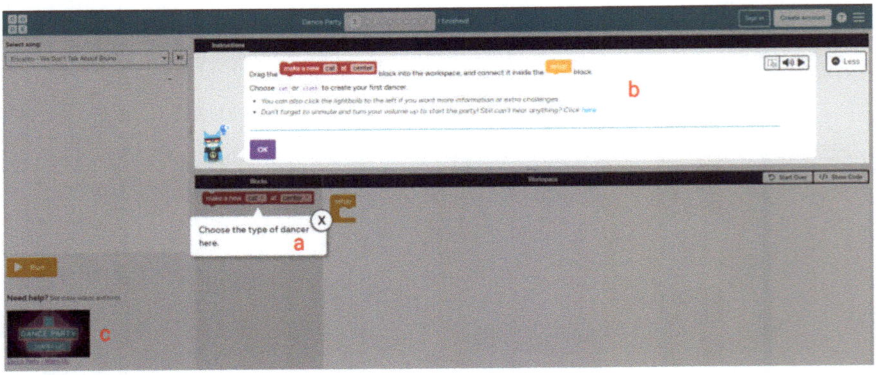

Fig. 3 The dance party support systems

6 Discussion

The presented Taxonomy of Support Methods for Serious Games for Programming (TSM-SGP) offers a comprehensive framework for identifying, classifying, and analysing support methods and systems in SGs about programming. It addresses a gap in the literature by providing a structured approach to the classification of the diverse range of support methods employed in educational games. The taxonomy allows researchers, educators, and game designers to systematically analyse and compare various support approaches by organizing methods into distinct categories. This structured framework enhances the understanding of existing support methods and provides a common language for discussing and evaluating these systems across various studies and applications. As the field of educational gaming continues to evolve, new support methods and technologies are likely to emerge. The taxonomy's flexible and expandable design allows for easy incorporation of these innovations. For instance, if augmented reality becomes a prevalent support method, it can be seamlessly added to the media category. Similarly, the potential emergence of hybrid AI systems as support sources can be accommodated within the existing framework. This adaptability ensures that TSM-SGP will remain relevant and useful as the field progresses. TSM-SGP, developed following the methodology proposed by Nickerson, Varshney, and Muntermann [12], can be considered useful as it demonstrates the following qualitative attributes:

- Conciseness, as it maintains a limited number of dimensions (six), each comprising a small set of characteristics (maximum 5).
- Robustness, as it contains the necessary categories and characteristics for accurately identifying and categorizing support methods in SGs.
- Comprehensiveness, as it is capable of classifying all known support methods within SGs for programming.

- Extendibility, as the taxonomy supports the addition of new categories and characteristics when new support methods appear, while also allowing for the removal of obsolete systems to maintain relevance during technological advancements.
- Explanatory, as it allows users to locate the position of a support system within the taxonomy and, in turn, derive its defining characteristics based on that classification.

While the TSM-SGP taxonomy represents a step forward in the classification and understanding of support methods in SGs for programming, it is important to acknowledge its limitations. Firstly, the taxonomy, although developed based on existing literature and expert insights, currently lacks empirical validation. This absence of rigorous testing in practical applications combined with the lack of experts' feedback may limit its immediate applicability across diverse educational contexts. Secondly, the subjective nature of categorization in research can lead to potential inconsistencies. Different researchers may interpret and categorize certain support methods differently, potentially resulting in disagreements or variations in implementation. This subjectivity could pose challenges when attempting to establish a standardized approach across the field.

Furthermore, the generalizability of the taxonomy across different domains and specific contexts may be limited. The effectiveness of support methods can vary significantly depending on factors such as subject matter, target audience, and specific learning objectives. What works well in one educational context may not be as effective in another, highlighting the need for further research to validate the taxonomy's applicability across diverse settings. Lastly, the rapidly evolving nature of technology and educational practices means that new support methods may emerge that are not yet accounted for in the current taxonomy. While the framework is designed to be flexible and expandable, ongoing updates and refinements will be necessary to ensure its continued relevance and comprehensiveness. Despite these limitations, future research holds significant potential to address these challenges through empirical validation, expanded applications in diverse educational contexts, and the incorporation of emerging technologies, ensuring the taxonomy's continued relevance and impact.

7 Conclusions

The TSM-SGP taxonomy offers a comprehensive framework for understanding and categorizing support methods and systems in educational games focused on programming. It establishes a structured approach to classifying various support methods and contributes significantly to the field of game-based learning and serious game design. The taxonomy enables a systematic way to organize and understand support methods and provides valuable insights into effective serious game design. The identified relationships between support methods and game properties contribute to the development of an integrated model of learning, advancing the understanding of how these educational tools can be optimized for maximum impact. While the current focus of this work is on programming in SGs, the framework established by TSM-SGP has the potential to serve as a foundation for exploring support methods in other educational domains. Its flexible and expandable nature allows for adaptation to diverse subject areas, potentially broadening its impact across the field of educational game design. Future research will empirically validate and refine the taxonomy to ensure its effectiveness and promote wider adoption in diverse educational contexts. In closing, the TSM-SGP taxonomy provides a valuable framework that advances the understanding and optimization of support methods in SGs for programming, enhancing educational games now and into the future.

Acknowledgement This work is part of a project that has received funding from University Of Macedonia Research Fund under the Basic Research 2023 funding programme.

References

1. Elder, C.D.: Serious Games. By Clark C. Abt. (New York: The Viking Press, Inc., 1970. Pp. 176. 4.95 paper.). Am. Polit. Sci. Rev. **65**(4), 1158–1159 (1971)
2. Susi, T., Johannesson, M., Backlund, P.: Serious games—an overview. https://www.diva-portal.org/smash/record.jsf?pid=diva2:2416 (2007). Accessed 5 December 2022
3. De Freitas, S.: Are games effective learning tools? A review of educational games. J. Educ. Technol. Soc. **21**(2), 74–84 (2018)
4. Mordor Intelligence: Serious games market analysis - Industry report - trends, size & share. https://www.mordorintelligence.com/industry-reports/serious-games-market (n.d.)
5. Wing, J.M.: Computational thinking. Commun. ACM. **49**(3), 33–35 (2006)
6. Belmar, H.: Review on the teaching of programming and computational thinking in the world. Front. Comput. Sci. **4**, 997222 (2022)
7. Bedwell, W.L., Pavlas, D., Heyne, K., Lazzara, E.H., Salas, E.: Toward a taxonomy linking game attributes to learning: an empirical study. Simul. Gaming. **43**(6), 729–760 (2012)
8. De Lope, R.P., Medina-Medina, N.: A comprehensive taxonomy for serious games. J. Educ. Comput. Res. **55**(5), 629–672 (2017)
9. Laine, T.H., Lindberg, R.S.: Designing engaging games for education: a systematic literature review on game motivators and design principles. IEEE Trans. Learn. Technol. **13**(4), 804–821 (2020)
10. Rato, D., Prada, R.: A taxonomy of social roles for agents in games. In: Entertainment Computing–ICEC 2021: 20th IFIP TC 14 International Conference, ICEC 2021, pp. 75–87. Springer, Cham (2021)
11. Hamrouni, A., Bendella, F.: A taxonomy of learner-player's emotions in serious games. Int. J. Serious Games. **10**(4), 17–32 (2023)
12. Nickerson, R.C., Varshney, U., Muntermann, J.: A method for taxonomy development and its application in information systems. Eur. J. Inf. Syst. **22**(3), 336–359 (2013)
13. Code.org.: AI for oceans: hour of code activity. Code.org. https://code.org/oceans (n.d.). Accessed 1 Jan 2025
14. Code.org. Dance party: hour of code activity. Code.org. https://code.org/dance (n.d.). Accessed 1 Jan 2025
15. Skudder, B., Luxton-Reilly, A.: Worked examples in computer science. Austral. Comput. Educ. **148**, 59–64 (2014)

16. Toukiloglou, P., Xinogalos, S.: Ingame worked examples support as an alternative to textual instructions in serious games about programming. J. Educ. Comput. Res. **60**(7), 1615–1636 (2022)
17. Soflano, M., Connolly, T.M., Hainey, T.: An application of adaptive games-based learning based on learning style to teach SQL. Comput. Educ. **86**, 192–211 (2015)
18. Qin, H. X., Jin, S., Gao, Z., Fan, M., & Hui, P.: CharacterMeet: supporting creative writers' entire story character construction processes through conversation with LLM-powered Chatbot avatars. In: Proceedings of the CHI Conference on Human Factors in Computing Systems, pp. 1–19 (2024)
19. OpenAI: Models - OpenAI API. https://platform.openai.com/docs/models/gpt-3-5 (n.d.). Accessed 19 May 2023
20. Thoppilan, R., De Freitas, D., Hall, J., Shazeer, N., Kulshreshtha, A., Cheng, H.-T., Jin, A., Bos, T., Baker, L., Du, Y.: Lamda: Language models for dialog applications. arXiv Preprint arXiv:2201.08239 (2022)
21. Toukiloglou, P., Xinogalos, S.: Effects of collaborative support on learning in serious games for programming. J. Educ. Comput. Res. (2024). https://doi.org/10.1177/07356331241296888
22. Toukiloglou, P., Xinogalos, S.: Adaptive support with working examples in serious games about programming. J. Educ. Comput. Res. **61**(4), 766–789 (2023)
23. Lindberg, R.S., Laine, T.H.: Formative evaluation of an adaptive game for engaging learners of programming concepts in K-12. Int. J. Serious Games. **5**(2), 3–24 (2018)

Serious Games Design for Higher Education: Aligning Bloom's Taxonomy, Cognitive Theory of Multimedia Learning, and Gamification for Effective Learning

Daphne Economou and Vassiliki Bouki

Abstract Serious games are increasingly prevalent in higher education, offering an engaging way to enhance student learning. But what are the key factors which determine whether a game embodies educational value or not? In this paper, we discuss the design of "Serious Games" in relation to their learning value and how we could address pedagogical needs through design. For this purpose, we use as a guide the theoretical foundation provided by Bloom's taxonomy, which describes the learning journey. Pedagogical needs are discussed in relation to the cognitive processes that appear in Bloom's taxonomy as well as in the Cognitive Theory of Multimedia Learning (CTML). After that, we discuss game design strategies that align with these needs. We propose a set of theory-driven design guidelines tailored to each level of Bloom's Taxonomy, illustrating their practical application through examples developed by the "Serious Games at Westminster" research group. These examples, spanning multiple disciplines, demonstrate how thoughtful design strategies can address diverse pedagogical needs and foster more effective learning experiences.

Keywords Serious games · Pedagogical needs · Bloom's taxonomy · Cognitive theory of multimedia learning · Gamification · Design guidelines

1 Introduction

Serious games have emerged as powerful tools in education, combining gamified elements with structured learning objectives to create engaging and effective learning experiences. Despite their growing adoption in higher education [1, 2], a fundamental

D. Economou (✉) · V. Bouki
University of Westminster, London, UK
e-mail: d.economou@westminster.ac.uk

V. Bouki
e-mail: boukiv@westminster.ac.uk

I. Kazanidis and A. Tsinakos (eds.), *Research on E-Learning and ICT in Education*,
https://doi.org/10.1007/978-3-031-99961-1_10

challenge remains: how to strike the right balance between educational rigor and engaging gameplay. The quest for high-quality learning games has spurred the development of various design frameworks and processes rooted in traditional instructional design models [3], which generally focus on either maintaining a balance of game elements [4], or adopting an iterative approach [5]. However, many serious games still lack a systematic framework that integrates educational goals with robust pedagogical theories, risking an overemphasis on entertainment at the cost of meaningful learning outcomes [6]. To address this shortcoming, it is crucial to align game mechanics with established educational principles, ensuring that serious games not only captivate learners but also enhance knowledge acquisition and cognitive skills [7].

In practice, game designers often prioritize entertainment while educators seek clear pedagogical alignment. A systematic design framework—one that merges Bloom's Taxonomy [8] with cognitive theories such as the Cognitive Theory of Multimedia Learning [9] and gamification principles—can reconcile these objectives. Bloom's Taxonomy provides a structured progression of cognitive objectives, ranging from foundational stages like remembering and understanding to higher-order skills such as evaluating and creating. Meanwhile, CTML offers strategies for optimizing multimedia learning by addressing cognitive constraints through principles like dual-channel processing, segmentation, and active learning. By weaving these theories into gamification elements—such as role-playing, storytelling, and feedback mechanisms—serious games can become both pedagogically impactful and engaging. Indeed, without such a framework, serious games risk becoming overly entertaining at the expense of meaningful, measurable learning outcomes [6].

This paper addresses that gap by proposing theory-driven guidelines that bridge established pedagogical concepts with practical game techniques, drawing on the work of the Serious Games at Westminster Research Group (SG@W). We illustrate the application of these guidelines through examples of serious games developed across various disciplines, including law, biomedical science, business, and political studies. These examples demonstrate how educational objectives can be achieved while preserving the essence of gameplay.

Additionally, we evaluate the effectiveness of one of our most comprehensive serious games, "Navigating Westminster," using analytics to examine student engagement and learning outcomes. Our findings highlight the potential of this design approach for tackling pedagogical challenges such as cognitive overload, short attention spans, and limited motivation, while fostering critical thinking and active learning. The paper concludes by discussing the broader implications of these findings and outlining future directions for advancing serious game design in education.

2 Addressing Educational Objectives

Bloom's taxonomy provides the theoretical foundation for structuring educational objectives. Originally introduced in 1956, it organizes learning goals into six categories: knowledge, comprehension, application, analysis, synthesis, and evaluation, progressing from simple to complex [10]. In 2001, the theory was revised to include: remember, understand, apply, analyze, evaluate, and create [8]. This updated version emphasizes learner-centered cognitive processes by using verbs (e.g., "remember" and "recall") rather than nouns, marking a shift from what teachers provide to how students learn.

Bloom's taxonomy also categorizes learning into three domains:

1. Cognitive: Mental skills like understanding and evaluation.
2. Affective: Emotions and attitudes.
3. Psychomotor: Physical tasks and abilities.

Universally applicable across disciplines, Bloom's taxonomy ensures our game design principles align with its educational foundation. In the following sections, we will address additional considerations in creating digital learning games, leading to examples of serious game design.

3 Understanding Learning and Its Role in Educational Games

Designing effective serious games requires a clear understanding of "learning" and its contributing factors. Disciplines such as Psychology, Pedagogy, and Neuroscience provide complementary insights into learning, which is broadly defined as a lasting change in knowledge or behavior. This change is evident when learners can explain and apply new knowledge, a process heavily reliant on memory. Existing knowledge can either facilitate or hinder the integration of new information, depending on how well it aligns with established cognitive structures [11].

Memory and cognitive capacity are at the heart of learning-focused design. While advanced technologies create exciting opportunities to integrate diverse information and build immersive simulations, they also present challenges. Excessive or conflicting content risks overwhelming learners, a phenomenon known as cognitive overload. Research highlights cognitive overload and declining attention spans as significant barriers in contemporary learning environments. Richard Mayer's Cognitive Theory of Multimedia Learning (CTML) addresses these challenges with a practical framework grounded in three foundational concepts:

1. Dual channels: The brain processes verbal and visual information separately, learning better from combined words and images than words alone.
2. Limited capacity: Cognitive resources are finite, so information must be carefully aligned across channels to avoid overload.

3. Active processing: Learning occurs when learners select relevant material, organize it meaningfully, and integrate it with long-term memory.

These principles underpin Mayer's 15 multimedia design guidelines (e.g., Coherence, Signaling, and Modality Principles), which offer strategies to reduce cognitive strain and enhance learner engagement. For example, presenting information in user-controlled segments or adopting conversational language can significantly improve learning outcomes [9].

Integrating CTML principles with Bloom's taxonomy further ensures that serious games support effective learning at every stage, from knowledge acquisition to higher-order thinking skills. Thoughtful alignment of these principles with game elements helps balance the cognitive demands of learning with the immersive, engaging qualities of gameplay.

In the following sections, we will review the gamification elements incorporated into our designs before delving into specific game examples.

4 Learning with Serious Games

While serious games aim to achieve educational objectives, they must also be fun, engaging, and fundamentally "games." Game-based learning leverages mechanics such as feedback loops, progression systems, and immersive narratives to sustain learner motivation and scaffold complexity [2, 12]. By embedding meaningful tasks within a structured flow of challenges, learners enter an optimal state of "flow" [13], ultimately boosting self-efficacy and deepening conceptual understanding [14]. In designing these experiences, mechanics, dynamics, and aesthetics play a pivotal role in mapping game elements to learning objectives [12], while connecting learning content with meaningful play ensures authenticity and relevance [15]. Balancing challenges and skill is crucial for sustaining engagement, as Csikszentmihalyi [13] illustrates in his work on flow and intrinsic motivation.

At the Serious Games at Westminster Research Group, we have developed games that are structured, goal-oriented, and sometimes competitive, depending on the mode of play [16–19]. These games feature specific rules, objectives, and engaging mechanisms, often including multiple levels and extended durations. They are not mere activities, but complete games designed to provide immersive learning experiences. To achieve this, we incorporate the following gamified elements:

1. Real-life scenarios: Situating learning in meaningful contexts, such as solving a real crime for law students or handling a domestic violence case.
2. Role-playing: Assigning learners roles (e.g., lawyer, judge, job applicant) with responsibilities and rewards tied to those roles.
3. Storytelling: Using narrative-driven contexts, enhanced with text, videos, pictures, and advanced techniques like comics or video scribe.
4. Direct manipulation: Allowing learners to create or make decisions by manipulating objects.

5. Interaction and object manipulation: Enabling direct interaction with learning materials, often resulting in screen changes and hands-on experiences.
6. Quizzes: Making quizzes more engaging by linking them to goals, such as revealing new images upon completion.
7. Points: Rewarding progress and achievements through points.
8. Level progression: Motivating learners to advance through game stages.
9. Branched scenarios: Introducing unpredictability with decision trees that lead to varied outcomes.
10. Visual and verbal feedback: Providing progress updates through messages or progress bars.
11. Embodiment: Using lifelike characters with gestures and expressions to interact with players.
12. Timed activities: Simulating real-life constraints to encourage focus and decision-making.

Notable examples of our work include a series of educational serious games designed to support undergraduate students at the University of Westminster across various disciplines. These include *Navigating Westminster*, an induction tool developed during the COVID-19 pandemic to facilitate online learning and orient new students; *Introduction to a Biomedical Lab*, which helps students at the School of Biomedical Science acquire essential health and safety knowledge and learn to use basic lab equipment before entering a real laboratory; a module for law students focused on mastering the *Theft Act*; *Business Concepts Tic Tac Toe*, a gamified activity that aids business students in understanding foundational business and computing concepts; and *Knowledge is Power*, a game designed to teach British Politics by enabling students to gain power as they expand their knowledge.

In the next section, we will explore specific examples and demonstrate how the discussed elements come together in designing effective educational serious games.

5 Theory-Driven Design Guidelines for Serious Games

One of the key challenges in designing educational serious games is the lack of design guidelines rooted in robust pedagogical foundations [6]. While serious games hold great potential for enhancing learning, their effectiveness often depends on the integration of sound educational principles with engaging gameplay mechanics [15]. Without clear, pedagogically informed guidelines, the design process risks focusing too heavily on entertainment at the expense of meaningful learning outcomes [20]. Addressing this gap requires aligning game design with established theories of learning.

In this section we propose a set of Design Guidelines (DGs) that address key pedagogical challenges in higher education by aligning with Bloom's taxonomy cognitive levels, incorporating principles of CTLT and gamification techniques. The guidelines are structured to ensure that educational objectives at each cognitive level

Table 1 Design guidelines for the creation of effective educational serious games combining bloom's taxonomy cognitive levels, CTML and gamification techniques

Bloom's taxonomy cognitive levels	Broader CTML principles	Gamification techniques	Design guideline
Remembering, understanding	Segmenting and simplification, dual-channel	Storytelling, interaction, branching scenarios	DG1: segment information with visual and narrative support
	Dual channel and personalization	Storytelling, interaction	DG2: use dual channels
Applying, analyzing, evaluating	Active learning and repetition	Quizzes, match pairs, feedback, level progression	DG3: reinforce recall with feedback and repetition
	Active learning	Direct manipulation	DG4: promote active learning through interaction and gamification
Evaluating, creating	Critical thinking	Reasoning challenges, points, feedback, progress, timed activities	DG5: foster critical thinking
	Personalization and contextual learning	Role-playing, real-world scenarios, branched scenarios, embodiment	DG6: engage emotionally and contextually

of Bloom's taxonomy are met while overcoming common obstacles such as cognitive overload, short attention spans, and lack of engagement. Table 1 outlines the design guidelines by linking Bloom's taxonomy cognitive levels, broader CTLT categories, and gamification techniques.

The subsequent section explores the applicability and effectiveness of these design guidelines through examples of serious games developed to meet educational needs in higher education, and specifically across different courses in multiple disciplines at the University of Westminster.

6 Implementing Design Guidelines in Educational Serious Games

This section presents the practical application of the proposed design guidelines through examples that address specific pedagogical challenges while ensuring educational value. Organized into three subsections aligned with Bloom's Taxonomy—"Understanding and Remembering," "Applying and Analyzing," and "Evaluating and Creating"—it demonstrates how the guidelines translate into effective resources, providing insights into their implementation and real-world impact.

The proposed guidelines are presented following the model in [21], which structures the presentation into four key components:

4. The *design guideline* itself,
5. The *motivation* underlying the guideline,
6. The *benefits* of applying the guideline, and.
7. An illustrative *example* of its practical use.

6.1 Design Guidelines Aligned to Understanding and Remembering

The proposed design guidelines presented in this section support the foundational stages of Bloom's Taxonomy, *Remembering* and *Understanding*, by facilitating recall, comprehension, and practical application of key information.

6.1.1 DG1: Segment Information with Visual and Narrative Support

1. *Motivation*: While higher education demands depth and comprehensiveness, excessive detail can overwhelm learners, leading to cognitive overload, short attention spans, and difficulties in processing and retaining knowledge [22]. Integrating storytelling with visual narratives mitigates these issues by organizing complex material into manageable segments that capture attention and provide meaningful context, ultimately making learning more accessible and engaging.
2. *Benefits*: Structuring information into smaller chunks follows the *Segmenting and Simplification Principles* of CTML, helping learners better process and remember key concepts. Visual aids—such as comics—further reduce cognitive load by replacing dense text with clear, digestible imagery. This approach also leverages the *Dual-Channel Processing Principle*, seamlessly blending verbal and visual elements while encouraging active problem-solving.
3. *Example*: "Navigating Westminster" is an educational resource that helps students become familiar with the campus, support roles, policies, and services through an engaging and interactive introduction to university life. Acknowledging the challenge of retaining extensive information, it combines games with structured information delivery to build foundational understanding and enhance both short- and long-term memory retention. Even if students don't remember every detail, they gain essential awareness of the university's organization and key concepts, enabling them to seek further information when needed. For instance, a mini-game (see Fig. 1) reduces cognitive load and improves retention by chunking information about student roles and services. The game features comic panels depicting Chao, an international student facing language barriers and seeking support. Through interactive storytelling, students learn about key university roles and whom to contact for specific challenges. By presenting information in a relatable and visually engaging context, the game simplifies complex

Fig. 1 Application of segmentation of information with dual channels efficiency to introduce faculty roles

institutional details into manageable segments, enhancing comprehension and recall. The scenario ends with Chao feeling empowered by the available support, illustrating how this approach fosters confidence and understanding in navigating university resources.

6.1.2 DG2: Use Dual Channels Effectively

1. *Motivation*: Engages both auditory and visual learners simultaneously, optimizing cognitive resources.
2. *Benefits*: Reduces cognitive load, enhances comprehension and retention by aligning verbal and visual information processing pathways (e.g., using synchronized narration and graphics rather than redundant text).
3. *Example*: To support students at the School of Biomedical Science, University of Westminster, master foundational skills like using a pipette, we developed a serious game utilizing flipped learning (see Fig. 2). This educational tool combines storytelling, dual-channel content, and interactivity to build confidence and competence. The module starts with an engaging introduction to the pipette,

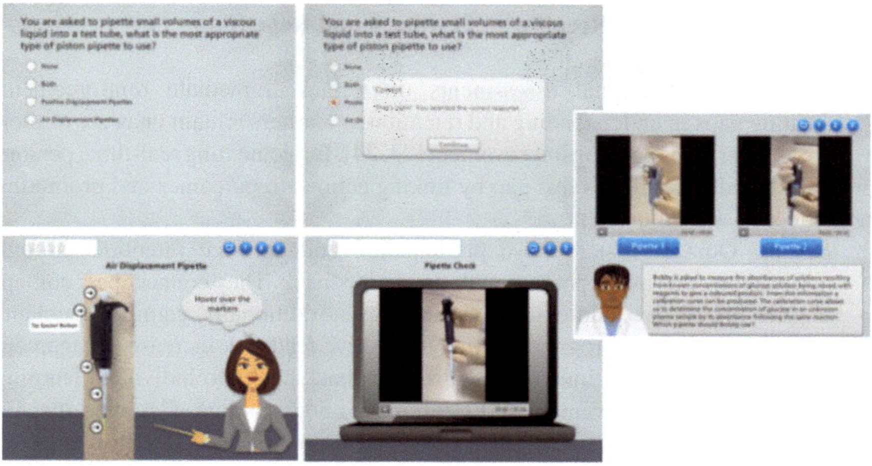

Fig. 2 Applying dual channel effectively in a biomedical science real-life scenario

featuring an interactive object that activates videos demonstrating proper pipetting techniques within meaningful tasks, such as preparing a sample for analysis. Students then participate in real-life biomedical lab scenarios, making choices and receiving personalized feedback on their decisions. This approach incorporates quizzes and feedback to address individual learning needs and aligns with Bloom's taxonomy by supporting both *Remembering* and *Understanding*.

The presented design guidelines collectively address the "Understanding" and "Remembering" levels of Bloom's Taxonomy by facilitating knowledge recall and comprehension through interactive and engaging methods. At the "Remembering" stage, activities like matching pairs and segmented information help students retrieve essential concepts, terms, and facts. At the "Understanding" stage, these guidelines enable learners to interpret, contextualize, and grasp the relevance of the information, often through visualizations, interactive explanations, or structured scenarios. This dual focus ensures that foundational cognitive processes are effectively supported, forming a strong basis for higher-order learning.

6.2 Design Guidelines Aligned to Applying and Analyzing

The proposed design guidelines presented in this section correspond to and enhance the mid-level stages of Bloom's Taxonomy, focusing on their role in deepening learning and engagement. By addressing the stages of "Comprehension," "Application," and "Analysis," these guidelines support learners in grasping core concepts, applying them to real-world scenarios, and critically examining relationships and patterns to develop higher-order thinking skills.

6.2.1 DG3: Reinforce Recall with Feedback and Repetition

1. *Motivation:* Traditional assessments often lack immediate reinforcement, creating gaps in understanding and retention as learners remain unaware of their progress or areas needing improvement [23, 24]. Implementing real-time, personalized feedback bridges this gap by linking actions to outcomes and promoting reflection and adjustment.
2. *Benefit:* Quizzes foster active participation and enhance retention through retrieval practice and repetition with timely feedback. They connect new content to prior knowledge, addressing learning gaps. Additionally, gamified elements such as points, level progression, and interactive feedback increase engagement and motivation. Real-time, personalized feedback caters to individual learning needs, guiding improvement and reinforcing understanding. Gamified mechanisms, like constructive feedback in drag-and-drop tasks, keep learners engaged and provide timely input that supports better comprehension and retention.
3. *Example:* We incorporated quizzes across various examples to help students familiarize themselves with available services and support (Fig. 1). These quizzes were integrated with real-life scenario simulations applying DG2 (Fig. 2), complemented by DG4 to promote active learning (Fig. 3), and DG6 to foster critical thinking through engaging activities (Fig. 4). This approach seamlessly combines multiple design guidelines to enhance comprehension, engagement, and higher-order cognitive skills.

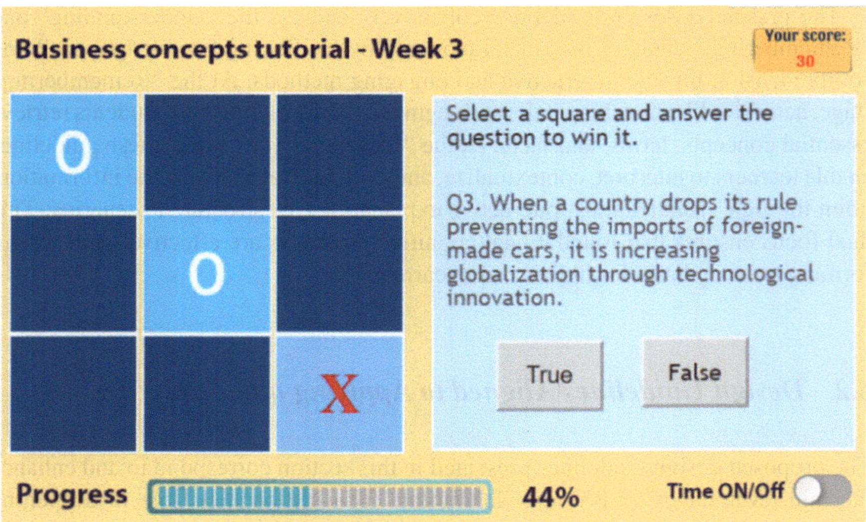

Fig. 3 Applying active learning through interaction and gamification

Fig. 4 Applying foster critical thinking

6.2.2 DG4: Promote Active Learning Through Interaction and Gamification

1. *Motivation*: Passive learning and low engagement lead to superficial understanding, limited retention, and hindered critical thinking [25]. The lack of active reasoning and meaningful interaction with materials necessitates strategies for deeper engagement and thoughtful learning. Additionally, time constraints utilize the gamification principle of urgency, motivating learners to think critically and act decisively.
2. *Benefits*: Active learning enhances attention and critical thinking through problem-solving tasks that foster analysis and hands-on engagement for deeper understanding. DG4 incorporates game mechanics to boost intrinsic motivation, encourage active reading, critical reasoning, and meaningful interaction with material. This approach aligns with self-determination theory, promoting sustained engagement and deeper learning.
3. *Example*: In a Level 4 Business Concepts module, students participate in a collaborative active learning activity to reinforce their theoretical knowledge. Structured as a tic-tac-toe game (Fig. 3), each square contains a question related to the studied concepts. When a student selects a square, they must reflect on and apply their theoretical understanding to answer the question. Correct answers earn points, fostering achievement and friendly competition, while incorrect answers receive immediate feedback to enhance comprehension. This game integrates interaction, gamification, and collaboration, promoting active engagement, critical thinking, and reflection. It may be played individually or as a whole-class activity.

The proposed guidelines align closely with the *Applying* and *Analyzing* stages of Bloom's Taxonomy, fostering active engagement and critical thinking. At the

Applying stage, students apply their knowledge in practical contexts, such as solving simulated scenarios or answering quiz questions based on learned concepts. This stage focuses on connecting abstract ideas to real-world applications, emphasizing the practical use of knowledge.

At the *Analyzing* stage, students evaluate and differentiate between elements like evidence and claims, assess trade-offs in decision-making (e.g., using gamified lifelines), and reflect on their reasoning. This encourages deeper comprehension by breaking down complex problems, exploring relationships, and developing metacognitive awareness.

By integrating theory with interactive, gamified activities, these guidelines transform traditional exercises into dynamic, cognitively engaging experiences, enhancing retention, comprehension, and meaningful learning outcomes.

6.3 Design Guidelines Aligned to Evaluating and Creating

The proposed design guidelines outlined in this section align with and enhance the higher-level stages of Bloom's Taxonomy, emphasizing their role in fostering advanced cognitive skills and deeper engagement. By addressing the stages of "Synthesis" and "Evaluation," these guidelines empower learners to integrate diverse concepts, create new ideas or solutions, and critically assess information and processes. This approach supports the development of higher-order thinking skills, enabling students to construct comprehensive understandings and make informed, reasoned judgments.

6.3.1 DG5: Foster Critical Thinking

1. *Motivation*: Critical thinking is essential in higher education for evaluating information and solving complex problems [26]. However, traditional methods often emphasize passive learning and memorization. DG5 addresses this gap by incorporating activities that promote analysis, reasoned judgments, and the exploration of alternative perspectives.
2. *Benefits*: This approach fosters deeper learning through reflective and analytical thinking, enhancing decision-making skills as students assess trade-offs and defend choices. Gamified elements such as points, bonuses, and penalties motivate learners by rewarding well-reasoned answers and creative problem-solving while promoting accountability. This dynamic environment not only enhances knowledge acquisition but also develops critical evaluation and application skills, preparing students for real-world challenges.
3. *Example*: To facilitate learning about British Politics, we developed the "*Knowledge is Power*" game (Fig. 4), a gamified activity that fosters critical thinking and higher-order skills. Players spin a fortune wheel to move along a

20-step path, answering questions to earn coins used to "buy" territories, simulating political power acquisition. Special squares introduce lucky cards that can double, deduct, or grant bonus coins. This combination of knowledge-based questions, risk-reward mechanics, and strategic decision-making encourages students to analyze choices, apply their understanding, and develop strategies, aligning with Bloom's Evaluating and Creating stages.

6.3.2 DG6: Engage Emotionally and Contextually

1. *Motivation*: Engaging students emotionally and contextually enhances higher-order thinking skills like Evaluating and Creating [27]. DG6 overcomes traditional methods' limitations by incorporating role-playing and real-world scenarios with relatable characters, fostering empathy and connection to drive deeper engagement and critical reflection.
2. *Benefits*: Role-playing and contextual scenarios immerse learners in evaluating situations, making decisions, and creating solutions, fostering critical thinking by encouraging multiple perspectives and innovative problem-solving. Emotional connections with embodied characters enhance retention and comprehension, making learning meaningful. DG6 personalizes and contextualizes abstract concepts, equipping learners with the skills and confidence to apply knowledge effectively in real-world settings.
3. *Example*: The application of this guideline using real-life scenario and relatable characters has been demonstrated in the examples presented in DG1 and DG2 (depicted in Figs. 1 and 2).

The proposed design guidelines collectively address the *Evaluating* and *Creating* stages of Bloom's Taxonomy by fostering higher-order cognitive skills through engaging, practical, and meaningful learning experiences. They bridge the gap between theory and practice, encouraging learners to critically analyze information, make thoughtful decisions, and develop creative solutions.

Gamified elements such as reasoning challenges, instant feedback, and interactive tasks make learning both engaging and effective by motivating students to weigh options, consider various perspectives, and sharpen decision-making skills. Role-playing and real-world scenarios further enrich this experience, giving learners hands-on opportunities to apply concepts in contexts that mirror actual challenges. This approach transforms abstract concepts into relatable, hands-on challenges, helping students deepen their understanding while building the confidence and skills they need to tackle real-world problems effectively. It's about making learning not just effective, but also meaningful and engaging.

Although each guideline presented in Sect. 6 addresses specific pedagogical needs—from preventing cognitive overload to fostering critical reflection—they can be combined for greater impact. For instance, DG1 (Segment Information) and DG6 (Emotional Engagement) together create manageable, context-rich modules that reduce cognitive strain while boosting motivation [1]. Careful integration is vital to avoid overwhelming learners, highlighting the need for adaptive design strategies that accommodate varying proficiency levels [14]. Finally, a user-centered, iterative process—encompassing prototype testing, feedback, and refinement—ensures that serious games remain pedagogically aligned while responding to technological constraints and evolving curricular goals.

6.4 Evaluation of the DG Based on Analytics

While evaluating each proposed design guideline is beyond the scope of this paper, we present an assessment of one of the most sophisticated implementations: the "Navigating Westminster" game for new students at the University of Westminster. This example offers valuable insights into the effectiveness of the design principles in practice as it applies all the presented guidelines discussed in this paper.

The "Navigating Westminster" game was closely monitored using Google Analytics during the first 5 months after its rollout (September 2020 to January 2021). During this period, which coincided with the pandemic, course leaders and personal tutors actively recommended the game to students. The results were highly promising: 1044 users participated, generating 1303 sessions. Players came from 48 countries and 207 cities, with 80.3% being new users and 19.7% returning users, indicating sustained interest and enjoyment of the game.

In total, the game garnered 10,550 screen views, with an average "Time on Screen" of 5 min and 20 s—highlighting strong engagement and active participation. The most visited sections included "Meet the Team," which aligns with DG1 and DG2 by addressing cognitive challenges in delivering information effectively and supporting the *Remembering* and *Understanding* stages of Bloom's Taxonomy. The "Roles" section aligns with DG3 and DG4, emphasizing *Applying* and *Analyzing* by reinforcing recall and encouraging active learning. Lastly, the "Support" section aligns with DG5 and DG6, promoting *Evaluating* and *Creating* through activities that foster critical thinking and emotional and contextual engagement.

These findings are encouraging, suggesting a positive response from students and validating the effectiveness of the proposed design guidelines in fostering engagement, understanding, and critical thinking. However, further testing and analysis are required to confirm these outcomes across different contexts and to refine the guidelines for broader application.

7 Conclusions and Future Directions

In this paper, we introduced a set of theory-driven guidelines for creating effective educational serious games that align Bloom's Taxonomy, the CTML, and gamification techniques. By addressing critical pedagogical challenges—including cognitive overload, short attention spans, and limited engagement—these guidelines foster higher-order thinking skills such as critical reasoning, problem-solving, and creative decision-making. Examples from the Serious Games at Westminster Research Group demonstrate the guidelines' versatility and applicability across various disciplines, while the analytics-based evaluation of *Navigating Westminster* underscores their potential to enhance engagement and learning outcomes.

Our work responds directly to gaps identified in prior literature. Lim et al. [28] call for an "intermediate framework" that balances gamification with cognitive context, warning that poorly chosen game elements can introduce irrelevant cognitive load. Our guidelines address this by integrating CTML to ensure cognitive alignment and by offering a structured approach for evaluating game mechanics in relation to learning goals. Similarly, Larsen et al. [29] emphasizes the need to ground game elements in learning theory to support both engagement and educational effectiveness. Our work extends this idea by combining multiple theoretical foundations and incorporating iterative design practices to maintain both educational integrity and player motivation.

Furthermore, we expand on Echeverría et al. [30], who noted the lack of explicit integration between games and the pedagogical process. While their framework highlights the need to balance the educational and ludic dimensions of classroom games, our guidelines go further by operationalizing this balance through Bloom's Taxonomy and CTML, and by supporting implementation through real-world case studies and educator collaboration. This evolution reflects the growing complexity of learning environments and the potential of serious games to address diverse educational needs.

Despite these promising findings, broader studies are needed to validate the guidelines across diverse contexts, courses, and learner demographics. Effective implementation also requires a user-centered, iterative approach: initial prototypes must be tested, refined based on feedback, and recalibrated to maintain both cognitive alignment and meaningful gameplay. As multiple guidelines are combined, designers and educators must remain aware of potential constraints or conflicts that might arise.

Future research will focus on long-term analyses of learning retention and the real-world application of knowledge gained through serious games. Integrating AI-driven adaptive learning shows promise for tailoring experiences to individual needs, and emerging technologies such as augmented and virtual reality may further enhance immersion, role-playing, and contextual learning scenarios. Ongoing collaboration with educators is essential to ensure that serious games remain aligned with curricula and responsive to learner needs.

While these efforts aim to refine and expand the guidelines, future research must also focus on developing a more systematic framework. Such a framework would provide a structured approach to designing and evaluating serious games, offering a cohesive methodology for balancing cognitive principles and resolving potential conflicts or restrictions. This approach addresses issues that have been is a need has been identified by designers and educators. By moving beyond standalone guidelines, this framework could guide the creation of more robust and adaptable educational tools.

Addressing these research directions will refine and broaden the scope of the guidelines, paving the way for impactful and inclusive applications of serious games in education. This work will continue to bridge the gap between pedagogy and game design, fostering innovative approaches to learning in the twenty-first century.

References

1. Bellotti, F., Ott, M., Arnab, S., Berta, R., de Freitas, S., Kiili, K., & De Gloria, A.: Designing serious games for education: from pedagogical principles to game mechanisms. In: Presented at the Proceedings of the 5th European Conference on Games Based Learning (2011)
2. Paas, F., Renkl, A., Sweller, J.: Cognitive load theory and instructional design: recent developments. Educ. Psychol. **38**, 1–4 (2003)
3. Kapp, K.M.: The Gamification of Learning and Instruction: Game-Based Methods and Strategies for Training and Education. Pfeiffer, Zürich (2012)
4. Silva, F.G.: Practical methodology for the design of educational serious games. Information. **11**, 14 (2019)
5. Arnab, S., Clarke, S.: Towards a trans-disciplinary methodology for a game-based intervention development process. Br. J. Educ. Technol. **48**, 279–312 (2017)
6. Einig, S.: Supporting students' learning: the use of formative online assessments. Acc. Educ. **22**, 425–444 (2013)
7. Van Staalduinen, J.-P., Freitas, D.: A game-based learning framework: linking game design and learning outcomes. In: Learning to Play: Exploring the Future of Education with Video Games, vol. 53, pp. 29–45. Peter Lang, New York (2011)
8. Prensky, M.: Digital game-based learning. Comput. Entertain. **1**, 21 (2003)
9. Johnson, J.: Designing with the Mind in Mind: Simple Guide to Understanding User Interface Design Guidelines. Morgan Kaufmann, San Francisco (2020)
10. Widmayer, S.A.: Schema theory: An introduction (2004)
11. Gerard, F.J., Mentzelopoulos, M., Economou, D., Khalish, Z., Ingram, J., & Ashley, E.: Work-in-progress—CrimOPS—gamified virtual simulations for authentic assessment in criminology. In: Presented at the 2022 8th International Conference of the Immersive Learning Research Network (iLRN) (2022)
12. Willingham, D.T.: Critical thinking: why is it so hard to teach? Arts Educ. Policy Rev. **109**, 21–32 (2008)
13. Khaleghi, A., Aghaei, Z., Mahdavi, M.A.: A gamification framework for cognitive assessment and cognitive training: qualitative study. JMIR Serious Games. **9**, e21900 (2021)
14. Ryan, R.M., Deci, E.L.: Self-Determination Theory. Springer, Cham (2024)
15. Ulum, Ö.G.: Is the revised bloom's taxonomy revisited in the EFL/ESL reading textbooks? OPUS J. Soc. Res. **19**, 170–177 (2022). https://doi.org/10.26466/opusjsr.1062878
16. Dimitriadou, A., Djafarova, N., Turetken, O., Verkuyl, M., Ferworn, A.: Challenges in serious game design and development: educators' experiences. Simul. Gaming. **52**, 132–152 (2021). https://doi.org/10.1177/1046878120944197
17. Bouki, V., & Economou, D.: Using serious games in higher education: Reclaiming the learning time. In: Presented at the 11th International Conference on Intelligent Environments (2015). https://doi.org/10.3233/978-1-61499-530-2-381
18. Massoudi, D., Koh, S., Hancock, P.J., Fung, L.: The effectiveness of usage of online multiple choice questions on student performance in introductory accounting. Issues Account. Educ. **32**, 1–17 (2017)
19. Bouki, V., Economou, D., Kathrani, P.: "Gamification" and legal education: A game based application for teaching university law students. Presented at the 2014 International Conference on Interactive Mobile Communication Technologies and Learning (IMCL2014) (2014)

20. Bouki, V., Economou, D., Angelopoulou, A.: Cognitive theory of multimedia learning and learning videos design: the redundancy principle. In: Presented at the Proceedings of the 29th ACM International Conference on Design of Communication. ACM, New York (2011)
21. Abd El-Sattar, H.K.: A new learning theory-based framework for combining flow state with game elements to promote engagement and learning in serious games. Inf. Sci. Lett. **12**, 2663–2677 (2023)
22. Hakim, M.W., Sari, D.M.M.: Practicing contextual teaching and learning approach to enhance students' higher order thinking skill on writing ability. ELSYA J. English Lang. Stud. **4**, 298–308 (2022)
23. Mayer, R.E.: The past, present, and future of the cognitive theory of multimedia learning. Educ. Psychol. Rev. **36**, 8 (2024)
24. Hunicke, R., LeBlanc, M., Zubek, R.: A formal approach to game design and game research. In: Presented at the Proceedings of the AAAI Workshop on Challenges in Game AI (2004)
25. Ávila-Pesántez, D., Rivera, L.A., Alban, M.S.: Approaches for serious game design: a systematic literature review. Comput. Educ. J. **8**, 11 (2017)
26. Csikszentmihalyi, M.: Intrinsic Motivation and Effective Teaching. Springer, Cham (2014)
27. Hmelo-Silver, C.E.: Problem-based learning: what and how do students learn? Educ. Psychol. Rev. **16**, 235–266 (2004)
28. Lim, T., Carvalho, M.B., Bellotti, F., Arnab, S., De Freitas, S., Louchart, S., Suttie, N., Berta, R., & De Gloria, A.: The LM-GM framework for serious games analysis (2015)
29. Larsen, T.M., Endo, B.H., Yee, A.T., Do, T., Lo, S.M.: Probing internal assumptions of the revised bloom's taxonomy. CBE Life Sci. Educ. **21**, ar66 (2022)
30. Echeverría, A., García-Campo, C., Nussbaum, M., Gil, F., Villalta, M., Améstica, M., Echeverría, S.: A framework for the design and integration of collaborative classroom games. Comput. Educ. **57**, 1127–1136 (2011)

Insights from Pre-Service Teachers Designing Location-Based Games

George Koutromanos

Abstract The aim of this study was to examine pre-service teachers' perceptions regarding the advantages of Location-Based Games (LBGs) and the factors affecting their use in teaching and learning. Data was collected using open-ended questions answered by 251 pre-service teachers who had previously designed and developed their own LBGs. The results indicated that pre-service teachers believe LBGs have specific learning benefits, in so far as they increase students' motivation, and improve their orientation, collaboration, and problem-solving skills. Moreover, they believe these games hold unique affordances such as giving students the ability to interact with the real environment and to combine both the digital and real world. Finally, they provide opportunities for exercise and physical activity. However, findings also reveal that pre-service teachers believe that a number of factors will hinder the use of LBGs in education. Some examples include student safety, teachers' lack of time, and the lack of digital infrastructure in schools. These results have various implications for the use of LBGs in education.

Keywords Location-based games · Pre-service teachers · Perceptions

1 Introduction

Over the past two decades, technological developments in mobile devices (i.e., tablets and smartphones), as well as telecommunications (i.e., 5G) have led to an emerging genre of games known as mobile Location-Based Games (LBGs). These games are also referred to as mobile-based games [1], urban games, street games [2], pervasive games [3], or location-based augmented reality games [4]. They are played outdoors in specific locations (e.g., city streets), built-up urban environments, as well as areas with special historical or cultural value [2].

G. Koutromanos (✉)
Department of Pedagogy and Primary Education, National and Kapodistrian University of Athens, Athens, Greece
e-mail: koutro@primedu.uoa.gr

I. Kazanidis and A. Tsinakos (eds.), *Research on E-Learning and ICT in Education*,
https://doi.org/10.1007/978-3-031-99961-1_11

LBGs combine the digital/virtual and real world. Through activities such as taking photos and recording videos, completing quizzes, interviewing passers-by, treasure hunts, and problem-solving, players must individually or collectively complete various tasks depending on the game's mission [5, 6]. A previous literature review by Avouris & Yianoutsou [2] separated LBGs into the following three categories: (a) ludic, (b) pedagogic, and (c) hybrid. Ludic games are played exclusively for pleasure and entertainment. Pedagogic games have a purely educational purpose. Hybrid games are designed for both entertainment and learning.

Previous and recent studies as well as literature reviews show that LBGs affect learning [1] and student motivation [7]. Moreover, research shows that students develop various skills (e.g., social and collaboration skills) [8] and improve players' physical activity [9]. Although these games bring added value to education, they have not been widely adopted by the education community. One possible reason is that the tools once available for game development were not user-friendly for teachers. Indeed, a literature review [10] revealed that most games used in research methodologies are developed mainly by researchers, study authors, or companies, rather than the educational community. Nonetheless, at present, the use of LBGs by teachers is more feasible thanks to several easy-to-use LBG authoring tools [6, 11] that do not require programming skills [10].

Although a significant number of studies have been conducted on the design [6] and effect of these games on learning [1, 12, 13], empirical findings on the experiences and beliefs of teachers involved in the development of such games remains limited [14]. This study aims to fill this gap with a sample of pre-service teachers who designed and played their own LBGs for the primary school level as part of their formative evaluation during an undergraduate course. The purpose of this study is to examine the following research questions: (a) What do pre-service teachers believe are the benefits of LBGs in primary education? (b) What do pre-service teachers perceive as the factors that affect the integration of these games in teaching and learning? and (c) What is their intention to design and use such games in the future as in-service teachers? The results of this study will help university undergraduate programs facilitate pre-service teachers' readiness to design and adopt LBGs in their teaching practices in the future.

This chapter continues with Sect. 2 presents indicative studies regarding teachers' beliefs about LBGs in education. The study's methodology is presented in Sect. 3, and the results of the data analysis are presented in Sect. 4. The chapter ends with a discussion of the results in Sect. 5, together with implications, limitations, and suggestions for further research.

2 Location-Based Games in Education

According to the literature, a research gap exists on LBG design by pre-service teachers, their perceptions on their benefits, and the factors that affect their use in teaching and learning. The few existing studies focus on in-service teachers' perceptions and can be separated into two categories. The first category includes studies whose main purpose was to examine the effect of LBGs on students' learning and as such teachers' perceptions of these games are included in the results. For example, one study conducted by Squire and Jan [15] examined the degree to which the location-based augmented reality (AR) game "Mad City Mystery," whose scenario was implemented on site at the University of Wisconsin-Madison campus near Lake Mendota, supported students in developing scientific argumentation skills. The teachers who accompanied their classrooms noted their students were excited about the game and the science and inquiry related to its scenario, particularly in relation to their local community. Four teachers in a study by Kamarainen et al. [16] expressed similar positive perceptions about a location-based AR game their students played which focused on a local pond environment. Students were encouraged to interact with the pond and their classmates, and the use of AR in the game prompted them to actively perform in the physical world, identify scientific practices, and motivated them to learn using a different approach. Moreover, the game provided students the opportunity for collaborative communication and problem-solving. Teachers, however, expressed concern over their own ability to manage the game's technology and devices while having to supervise travel to the pond environment.

The second category includes studies in which teachers express their perceptions of LBGs (e.g., usability, recommendation for further improvement, identification of problems) within the context of a formative evaluation of an LBG [17, 18] or assess their overall acceptance in a teaching environment [14]. An example on usability is provided in a study by Meishar-Tal and Ronen [18], in which 26 teachers played the location-based AR game "EduPARK" during a 4-h workshop conducted outdoors, following which they gave a positive evaluation on its user friendliness and educational value.

Likewise, in a study by Marques and Pombo [18], 35 teachers played LBG "my hometown" using Treasure-HIT. The results showed that the sample expressed favorable opinions of the game in terms of ease of use, enjoyment, motivation, cooperation, and knowledge acquisition. Furthermore, many teachers considered the game to be easy for their students' aptitude level. In a study by Koutromanos et al. [19] on the development and formative evaluation of an augmented reality LBG aimed at teaching the local history of an island, the two teachers who used the game with their students mentioned challenges over organizing their students into groups due to the large number of tourists present at the site. Additionally, they cited Internet connectivity issues. Setting out to evaluate an augmented reality LBG, Costa et al. [20] found that the teachers in the sample believed the game promotes students' interest in learning and is appropriate for a primary school curriculum.

An example of LBG acceptance in teaching was confirmed in a recent study by Karanasios and Papastergiou [14]. In their study, an open-source platform called "YOUth Go" was designed and developed to create LBGs that require players' physical activity. Initially, 84 in-service teachers created their own single-player or multiplayer educational LBGs and subsequently evaluated the platform's usability. They assessed their acceptance of the platform through the Unified Theory of Acceptance and Use of Technology (UTAUT) model. The results showed that the teachers had positive perceptions of all the variables in the UTAT and a positive intention to use "YOUth Go" in their teaching.

3 Methodology

3.1 Sample

Participants in this study included 251 undergraduate students (pre-service teachers) from the Department of Pedagogy and Primary Education of the National and Kapodistrian University of Athens, Greece. Students participated in the study as part of a compulsory course which is ICT in Education. This course aims to educate students on the pedagogical use of ICT in primary education and provides them with the appropriate digital skills to integrate ICT across all subjects in their future teaching. All students were informed that their responses were anonymous and signed a consent form to participate. Of the 251 participants, 216 (86.1%) were women and 35 (13.9%) were men, while the average age was 20.3 years. None had previously developed an LBG or any other kind of digital game.

3.2 Data Collection

Data was collected using open-ended questions, the responses of which were uploaded on a Word file to the course's Learning Management System along with a description of the game (i.e., subject, learning objectives, content of the games, and an indicative method of implementation). In addition, links to the games developed by the pre-service teachers were provided. The questions pre-service teachers were requested to answer included: (1) What do you believe are the advantages of the LBG you developed for primary school students? (2) What do you believe are the challenges that a teacher might face when designing and implementing an LBG in their teaching and on their students' learning? (3) Do you intend to design and use LBGs in your teaching in the future?

3.3 The Procedure

The study was conducted within the context of an undergraduate compulsory course called ICT in Education. It lasted approximately one and a half months (May–June 2024) and consisted of four phases. In Phase 1 (3 h), students were informed about mobile learning and digital games in education. In Phase 2 (3 h), groups of 24 students (1 person per computer) familiarized themselves with the environment and characteristics of Actionbound (https://en.actionbound.com/) during the lab component of the course, which took place in the university's Computer Laboratories. Actionbound is an easy-to-use platform employed in previous studies to develop LBGs [11, 21]. Through this platform, users can develop LBGs using game elements and tools such as voice recordings, videos, pictures, GPS locations, missions, directions, maps, quizzes, QR codes, and tournaments. In Phase 3, each student was asked to design an LBG appropriate for the primary school curriculum using the Actionbound platform in her/his free time over three weeks. In addition, it was requested that the pedagogical design be founded on teaching strategies from well-known learning theories that the students had already been taught within the context of the course and their studies in general (e.g., social constructivism, situated learning, etc.). They were free to choose the location of the LBG and its subject (Fig. 1).

Moreover, before submitting, they were asked to play their game at the site for which they had designed it to check for possible problems (e.g., spots not covered by the mobile device network signal, GPS divergence), as well as other factors that could hinder the implementation of the game. In Phase 4, each student needed to hand in a written description of the game, answer the three research questions that were previously mentioned, and upload their Word file to the course Learning Management System.

3.4 Data Analysis

Qualitative data from pre-service teachers' responses were analyzed using thematic content analysis. This was performed by the study's author, along with an independent researcher who is an expert in digital technologies in education. Initially, the data were coded to identify categories and sub-categories for the study's research questions. Having discussed similarities and differences found in the categories and sub-categories, the two researchers subsequently reached a consensus on a set of common codes. The codes were then categorized and possible relationships between the categories were examined. Finally, indicative responses for each emerging category and sub-category were selected [22].

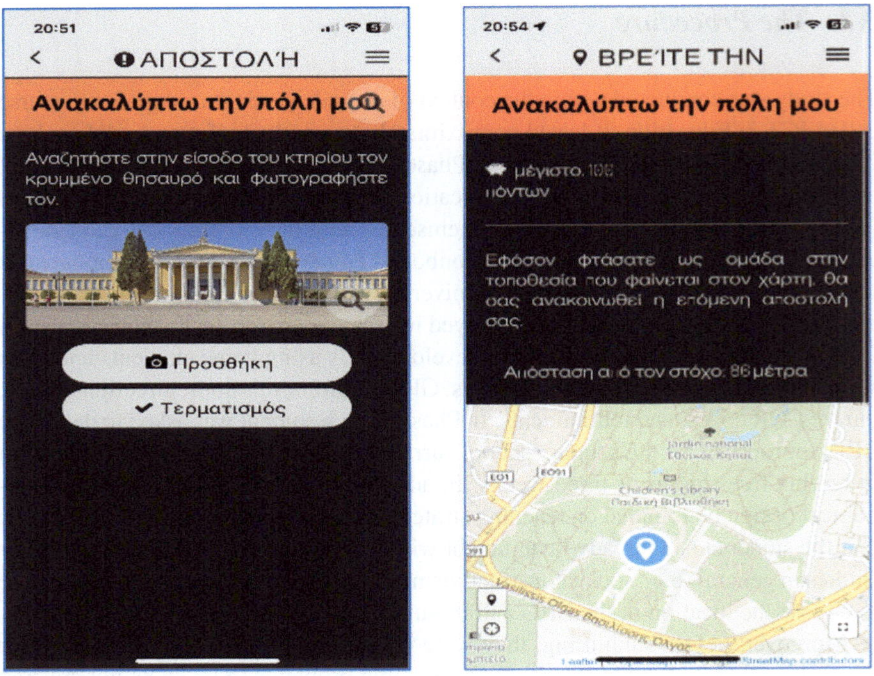

Fig. 1 Indicative examples of screenshots from the location-based game developed by a pre-service teacher. In the screenshot on the left, players were asked to photograph a specific object near the building shown in the image. In the screenshot on the right, players were asked to go to a particular location to receive a specific mission related to the game

4 Results

The 251 LBGs the students developed were examined prior to the analysis of their responses. Of these, 140 (55.8%) included content and activities about History, 48 (19.1%) on Natural Sciences, 23 (9.2%) on Mathematics and 19 (7.6%) on Language and 21 (8.4%) LBGs contained cross-curricular content.

Table 1 summarizes the results from the analysis of 255 pre-service teacher responses (henceforth referred to as PS-T1, PS-T2, … PS-T255). As shown below, the advantages of mobile LBGs in primary education can be classified into three categories: (a) learning benefits, (b) affordances, and (c) opportunities for exercise and physical activity. Furthermore, as Table 1 shows, several barriers that affect the use of LBGs were identified using the data. The sub-sections that follow provide an analytical presentation of these results.

Table 1 Factors regarding the use of mobile LBGs in primary education

Categories	Factor	N
Learning benefits	Makes learning more enjoyable and interesting	210
	Increases learning motivation	190
	Enhances skills	
	– Orientation	176
	– Collaboration	169
	– Problem solving	150
Affordances	Interaction with the real environment	201
	Combination of the digital and real environments	180
Opportunities for exercise and physical activity	Improves physical fitness	163
Challenges and concerns	Student safety	223
	Teachers	
	– Lack of time to design and implement the LBG	178
	– Difficulties coordinating/organizing students in an outdoor setting	154
	Digital infrastructure	
	– Internet connectivity issues	160
	– Inadequate number of mobile devices	120
	Curriculum-related restrictions	111
	Resistance to change	68
	Accessibility for students with mobility issues at locations where LBGs are played	56
	Weather conditions	45

4.1 Learning Benefits of Mobile Location-Based Games

The majority of sampled pre-service teachers believe that LBGs possess significant learning benefits for students. First, they believe LBGs make learning more enjoyable and interesting (N = 210), as evidenced by the following quotes:

> A sense of adventure and treasure-hunting is created, and this makes it more interesting and enjoyable for students (PS-T 44).
>
> Children don't see learning as an evaluation, or a stressful, competitive activity, but rather as an interesting and enjoyable game that provides them with new skills… (PS-T 94).
>
> …participation in learning is encouraged in a more pleasant and enjoyable manner among all students compared to learning in the classroom (PS-T 188).
>
> The interactive, multimedia activities and treasure hunts within a natural environment improve student participation and enable them to experience education as an enjoyable and interesting process (PS-T 68).

Moreover, the pre-service teachers (N = 190) believe that playing LBGs increases students' motivation to learn. As quoted:

> I believe the most important factor is increased motivation (PS-T 76).
> ...I believe students will be more motivated to learn through the game I created compared to a standard school classroom ... (PS-T 58).
> I think what's often missing in schools is student motivation. I think that LBGs encourage students' motivation to learn so long as it takes place in a real environment... (PS-T 158).

Participants also mentioned that LBGs support the development of various skills. One such skill is orientation, considering that students need to carry out certain tasks in specific environments. Other skills such as conducting research and collecting data are also encouraged (N = 176). For example:

> Students' orientation is developed with the help of maps as they learn to distinguish roads and locations (PS-T 49).
> ...students need to search for information and data, and go to specific site locations. Consequently, they develop their spatial understanding and identification of a specific location (PS-T 32).
> The game helps the development of students' spatial understanding because they explore the area and try to navigate it (PS-T 119).
> ...by researching buildings, statues, as well as information on the history of a place and exploring it based on the game's instructions, students learn to orient themselves across various places and locations (PS-T 71).

Another perceived learning benefit is the enhanced ability to collaborate (N = 169). Indicative responses are presented below:

> The treasure-hunt activities and decision-making process encourage collaboration and teamwork, fostering open communication and discussion among children in the real world (PS-T 55).
> ...one skill that can be improved through this game is collaboration between classmates since the children are asked to become part of a group and work together to complete a task (PS-T 93).
> An additional advantage the game offers is that it promotes collaboration between students by enhancing their social/teamwork skills and the exchange of ideas (PS-T 172).
> This specific educational game promotes collaboration thanks to the tasks and activities it features, which require collaboration between team members for successful completion (PS-T 196).

Another benefit mentioned is improved problem solving (N = 150). As the following pre-service teachers noted:

> There is an improvement in problem-solving skills. Problem solving in real time and space requires adaptation and creativity, therefore students learn to overcome obstacles and adjust to new real-world problem – solving situations (PS-T 2).
> It strengthens the ability to solve problems because students need to follow a specific set of interrelated steps to complete the game (PS-T 38).
> Students are encouraged to solve problems while carrying out tasks. Therefore, they think critically and methodically, while also seeking solutions through observation and the analysis of information available to them (PS-T 102).
> This game enhances students' problem-solving skills because they participate in a process of investigation, research, and contemplation to achieve the game's mission, which is to successfully complete the assigned tasks (PS-T 131).

4.2 *Affordances of Mobile Location-Based Games*

The pre-service teacher response analysis showed that LBGs have two unique affordances. The first is the interaction these games have with the real environment (N = 201). The second is the combination of the digital and real environments (N = 180). The following are indicative responses given by pre-service teachers supporting the first affordance:

> There is direct interaction between students and the real environment since they come into contact with real theme-based activities, in contrast to digital games, which focus on imaginary activities (PS-T 4).
>
> The game's real experiences provide students with the opportunity to interact with their environment in a natural way. This interaction enhances comprehension and learning as students experience the lesson's content in real conditions (PS-T 19).
>
> The most important factor is interaction with the real world. Students are asked to move around in the natural environment, explore it, observe things that they may have been unaware of until now, and seek out information for themselves. Thus, learning becomes experiential and more tangible for students (PS-T 15).
>
> ...the students find themselves in a real environment and can see everything displayed live in the game's plot, thus engaging all their senses (PS-T 1).
>
> ...this application encourages students to go to outdoor spaces, discover information, as well as experiment through real activities. Consequently, they become creative researchers (PS-T 251).

The following examples demonstrate how pre-service teachers see LBGs combining the digital and real worlds.

> This game combines contact with the real and digital environments, providing students the opportunity to do something creative... (PS-T 14).
>
> ...I believe the game I created is unique because it incorporates technology with empirical learning in a real environment resulting in more effective teaching (PS-T 25).
>
> ... the combined use of digital technology and real elements transforms games played in a physical environment into a pleasant activity, and more so into a powerful learning tool that exceeds the boundaries of traditional digital games (PS-T 135).

4.3 *Opportunities for Exercise and Physical Activity*

A significant number of pre-service teachers (N = 163) believe that students can improve their physical fitness by playing LBGs. Four quotes in this regard include:

> Students participate in physical activity thus improving their physical health. Playing the game outdoors reduces the time students spend seated in front of a screen (PS-T 79).
>
> Digital games limit movement and interaction to a virtual world, while location -based games improve students' actual mental and physical health (PS-T 21).
>
> It encourages physical activity, since students must walk, run, and move to complete their tasks (PS-T 154).
>
> A location-based game encourages physical movement since students walk around and explore their surrounding environment, thus promoting a healthy lifestyle (PS-T 142).

4.4 Challenges and Concerns

One of the most significant issues identified in the analysis of pre-service teachers' responses was student safety (N = 223). The main concerns included traffic-related injuries, outdoor injuries, as well as students getting lost in crowded places. Here are some indicative responses.

> …to me student safety is the most important factor as public spaces contain dangers that are often hard to manage (PS-T 11).
>
> When students are outdoors, their safety poses a significant concern. …students mustn't get lost or be exposed to dangers during the game (PS-T 162).
>
> One basic disadvantage in the use of this game is that it creates a level of risk since children are asked to go outdoors and cross streets (PS-T 39).
>
> … Accidents or losing students are very likely to happen in outdoor settings (PS-T 43).

Challenges in implementing LBGs are also related to teachers. One such challenge is the amount of time required for teachers to develop, plan, and implement an LBG on a specific site (N = 178). In this respect, pre-service teachers 5, 146, 198, and 212 were quoted as saying:

> The 'Treasure Hunt' that I created requires quite a bit of teacher preparation time to implement it in real conditions (PS-T 5).
>
> Creating such a game demands more time from the teacher because they need to design and deliver it at a specific location, while also providing the necessary support and appropriate feedback to students, as this type of learning is new and unfamiliar to them (PS-T 146).
>
> … requires greater preparation and coordination from teachers, considering they need to design the game, plan the activities in a natural environment, as well as visit the site beforehand prior to taking the students (PS-T 198).
>
> …limited teacher time. Planning to visit an archaeological site to carry out the game's activities requires additional time compared to a classroom lesson (PS-T 212).

Additionally, more than half the participants (N = 154) mentioned that student coordination and organization in outdoor settings is another significant factor that will affect the implementation of an LBG. The following pre-service teachers note:

> Controlling student behavior outdoors can become difficult for a teacher, especially if the group is large... (PS-T 222).
>
> … difficulties arise when having to play the game since dividing students into teams can be challenging for teachers; especially if the class is large or the students present different learning needs (PS-T 123).
>
> Coordinating large student groups in a real environment requires careful planning, organization, and proper supervision from teachers (PS-T 137).
>
> …I believe that one problem teachers will encounter is having to organize the students. Since the children may lack enthusiasm for this new way of learning, they may lack focus and display behaviors that can cause problems… (PS-T 177).

Pre-service teachers (N = 160) also stated that poor Internet connectivity is a significant obstacle in implementing LBGs. The following excerpts demonstrate this prominently.

> …various technical issues such as loss of Internet connectivity in an outdoor environment, or inaccurate location detection by the GPS may occur (PS-T 142).

> The game requires a stable Internet connection, which might be a problem in areas where there is poor coverage or no available Wi-Fi signal (PS-T 168).

> Most of the game's functions require an Internet connection, which might not be possible in all the outdoor areas students may be exploring (PS-T 208).

> Possible technical issues, such as the application's connectivity or operation may disrupt the smooth flow of the game (PS-T 243).

The lack of a sufficient number of mobile devices in school represents another barrier according to 120 pre-service teachers. Examples of this disadvantage are evident in the responses of pre-service teachers 148, 192, and 206 cited below.

> The basic prerequisite for its success is that the number of tablets available is equal to the number of students (PS-T 148).

> Furthermore, schools might be missing updated tools or equipment (tablets) to accommodate all students, without which the game cannot be used (PS-T 192).

> The school might not own the necessary number of tablets (PS-T 206).

Additionally, several pre-service teachers (N = 111) mentioned that the curriculum might pose another barrier toward using LBGs in teaching and learning. The following excerpts describe this barrier.

> It might be difficult to incorporate the game into the curriculum, especially if its implementation is lengthy or there are other educational obligations that must be met (PS-T 38).

> One significant concern I have regarding the use of this game is possible difficulties incorporating it into the existing curriculum, which may occur if its aims are not achieved in an effective manner (PS-T 171).

> …it is important to ensure that the game's educational aims align with the general curriculum and that the students receive the greatest possible benefits from the activity (PS-T 201).

> … the activities require a significant amount of time and cannot be completed in a single hour or day. Therefore, a generous amount of teaching time taken from the curriculum for this specific topic will be required (PS-T 211).

Based on the analysis of the sample's responses, it was found that resistance to change (N = 68) from some parents and teachers around this type of learning outside the classroom might hinder the implementation of LBGs. For instance, pre-service teachers expressed the following viewpoints:

> …some parents might not want their children to participate for various reasons and may not give their consent (PS-T 183).

> A difficulty that may arise in the future with the implementation of this game is that parents or teachers may not be willing to go to the park to use the application (PS-T 239).

> If I decide to present this application to my students in the future, I believe that I will encounter certain difficulties, beginning with the fact that this type of digital game is a fairly innovative idea that is not at all common in education in our country. Therefore, it is possible that it might face scrutiny – more so by parents and colleagues than by students – because it runs contrary to the traditional educational model that they are used to (PS-T 51).

> Finally, the desired cooperation and enthusiasm from other teachers may be missing due to the amount of time they would need to dedicate. Nor is it certain that parents would consent to such activities because they would feel that their children would lose time from other subjects that are important to them (PS-T 115).

Some pre-service teachers (N = 56) cited the issue of equal access for students with disabilities. For example:

> The location selected for the game I designed may present limited accessibility for students with mobility issues or disabilities, which may prevent their participation in places that are not accessible (PS-T 173).

> Students with mobility issues might have a difficult time participating in the activities. Special adjustments must be made to ensure accessibility and equal participation for all students (PS-T 219).

> One of the game's primary shortcomings is that it excludes people with mobility issues or other disabilities because it requires constant movement across various locations in the city (PS-T 247).

Furthermore, another significant factor mentioned by the pre-service teachers (N = 45) that will affect students when playing mobile LBGs is weather conditions. The following remarks by PS-T 29 and PS-T 125 are worth noting:

> The weather could adversely affect the implementation of the game. Rain and cold, or too much heat can make outdoor activities difficult and dangerous (PS-T 29).

> …the game requires physical presence and movement in outdoor locations, which is something that might not always be possible due to unstable weather conditions (a thunderstorm, intense heat, etc.) (PS-T 125).

4.5 Pre-Service Teachers' Intention to Use Location-Based Games in their Future Teaching

Of the 251 pre-service teachers, the majority (N = 231, 92%) stated they intend to design and use LBGs in their teaching once they are appointed. Only 20 (8%) rejected the possibility of future use. Some reasons are similar to those mentioned in the previous section: lack of time, difficulty associated with coordinating students in outdoor areas, and technical problems. As two pre-service teachers note respectively "*…perhaps this type of game doesn't suit me*" (PS-T 111), and "*I believe I'll be able to utilize my time on other more creative activities inside the classroom*" (PS-T 199).

5 Discussion and Conclusion

The purpose of this qualitative research was to investigate pre-service teachers' perceptions regarding the advantages of LBGs in primary education, as well as the factors that may affect the integration of these games in teaching and learning. The study found that pre-service teachers have positive perceptions of this genre of games and intend to use them in their future teaching. Firstly, they believe that LBGs possess specific learning benefits (i.e., they make learning more enjoyable and interesting, increase motivation, enhance orientation, collaboration, and problem-solving skills). These findings are in line with previous research about the advantages and learning benefits of LBGs in education [16, 18]. Secondly, the pre-service teachers believe LBGs present two unique affordances: interaction with the real environment and the combination of the digital and physical worlds. Although these two affordances have been identified in previous studies, they have not been evidenced prominently [16, 20]. Thirdly, pre-service teachers believe playing LBGs offers students opportunities for exercise and physical activity. This finding was also identified in a previous study that used the LBG Pokémon GO, which included players of various ages in its sample [4].

Aside from the benefits, the study also revealed several challenges and concerns regarding LBGs in teaching and learning. Consistent with past research, this study found that student safety when playing LBGs outdoors presents a significant barrier [19]. Two additional findings from the study, based on the sample, highlight factors influencing the development and implementation of LBGs in primary education: teachers' limited time and challenges in coordinating and organizing students in an outdoor environment. Previous studies also highlighted the importance of these factors, likely to affect teachers' design and use of the games [14, 16]. Another factor that prevents the use of LBGs is digital infrastructure—particularly Internet connectivity issues and lacking mobile devices in schools. This factor was also highlighted in previous LBGs studies [12, 19].

Pre-service teachers also believe that an inflexible curriculum poses a barrier to integrating LBGs in teaching. This difficulty was cited by teachers in previous studies on mobile learning applications—particularly on the use of AR [23]. Additionally, the pre-service teachers expressed their concern regarding possible reactions by other teachers and parents against the use of these games. The sample also mentioned restricted access for students with mobility issues to locations where these games are played as a factor that negatively affects their use. Finally, they cited weather conditions. In general, these factors are not commonly mentioned in much of the existing literature that sets out to examine the perceptions of pre-service and in-service teachers on the use of LBGs in education.

To the best of our knowledge, this study is likely one of the first to examine the perceptions of pre-service teachers on LBGs. The results enrich existing research with data on a teacher profile that is willing to design and develop their own LBGs. The results are encouraging regarding the future use of LBGs in education considering that pre-service teachers believe they offer numerous advantages to students. Therefore, university undergraduate education programs for pre-service teachers could encourage positive perceptions of LBGs by instructing them to design their own games. Regarding the factors identified as affecting the use of LBGs, these can be resolved in collaboration with the school community, as well as parents. For example, a possible solution to the issue of safety or planning in outdoor locations is for students to be accompanied by their parents or other teachers and/or volunteers from the school community. Moreover, to ensure that teachers implement LBGs in their teaching, the educational policies of a country should focus on equipping schools with the necessary mobile technology devices, as well as making curricula more flexible to incorporate these games. Finally, software companies could provide easy-to-use LBG development platforms featuring a variety of capabilities to create activities and content that favor the development of LBGs based on the advantages identified in this study.

Future research should consider the limitations of a convenient sample used in this study. Moreover, studies could examine the perceptions of pre-service teachers according to the genre of the game developed (e.g., adventure, treasure hunt, problem solving) or the pedagogical design adopted. Finally, another study could examine pre-service teachers' perceptions after they use these games within the context of the student-teaching in the classroom that they perform as part of their studies.

References

1. Huizenga, J., Admiraal, W., Ten Dam, G., Voogt, J.: Mobile game-based learning in secondary education: students' immersion, game activities, team performance and learning outcomes. Comput. Hum. Behav. **99**, 137–143 (2019). https://doi.org/10.1016/j.chb.2019.05.020
2. Avouris, N.M., Yiannoutsou, N.: A review of mobile location-based games for learning across physical and virtual spaces. J. Univ. Comput. Sci. **18**(15), 2120–2142 (2012)
3. Kasapakis, V., Gavalas, D.: Pervasive gaming: status, trends and design principles. J. Netw. Comput. Appl. **55**, 213–236 (2015). https://doi.org/10.1016/j.jnca.2015.05.009
4. Alha, K., Koskinen, E., Paavilainen, J., Hamari, J.: Why do people play location-based augmented reality games: a study on Pokémon GO. Comput. Hum. Behav. **93**, 114–122 (2019). https://doi.org/10.1016/j.chb.2018.12.008
5. Giannakas, F., Kambourakis, G., Papasalouros, A., Gritzalis, S.: A critical review of 13 years of mobile game-based learning. Educ. Tech Res. Dev. **66**, 341–384 (2018). https://doi.org/10.1007/s11423-017-9552-z
6. Xanthopoulos, S., Xinogalos, S.: An overview of location-based game authoring tools for education. In: Auer, M., Tsiatsos, T. (eds.) Interactive Mobile Communication Technologies and Learning. IMCL 2017. Advances in Intelligent Systems and Computing, vol. 725, pp. 243–255. Springer, Cham (2018). https://doi.org/10.1007/978-3-319-75175-7_21
7. Koutromanos, G., Sofos, A., Avraamidou, L.: The use of augmented reality games in education: a review of the literature. Educ. Media Int. **52**(4), 253–271 (2015). https://doi.org/10.1080/09523987.2015.1125988
8. Laato, S., Islam, A.N., Laine, T.H.: Did location-based games motivate players to socialize during COVID-19? Telematics Inform. **54**, 101458 (2020). https://doi.org/10.1016/j.tele.2020.101458
9. Khamzina, M., Parab, K.V., An, R., Bullard, T., Grigsby-Toussaint, D.S.: Impact of Pokémon go on physical activity: a systematic review and meta-analysis. Am. J. Prev. Med. **58**(2), 270–282 (2020). https://doi.org/10.1016/j.amepre.2019.09.005
10. Ribeiro, F.R., Silva, A., Silva, A.P., Metrôlho, J.: Literature review of location-based mobile games in education: challenges, impacts and opportunities. Informatics. **8**(3), 43 (2021). https://doi.org/10.3390/informatics8030043
11. Skaraki, E., Kolokotronis, F.: The use of the mobile application Actionbound in the teaching of local history. Adv. Mobile Learn. Educ. Res. **2**(2), 421–426 (2022). https://doi.org/10.25082/AMLER.2022.02.010

12. Admiraal, W., Huizenga, J., Akkerman, S., Ten Dam, G.: The concept of flow in collaborative game-based learning. Comput. Hum. Behav. **27**(3), 1185–1194 (2011). https://doi.org/10.1016/j.chb.2010.12.013
13. Kyza, E.A., Georgiou, Y.: Scaffolding augmented reality inquiry learning: the design and investigation of the TraceReaders location-based, augmented reality platform. Interact. Learn. Environ. **27**(2), 211–225 (2019). https://doi.org/10.1080/10494820.2018.1458039
14. Karanasios, S., Papastergiou, M.: In-service school teachers' evaluation of YOUth go, a platform for easily creating educational location-based games that require players' physical activity. Educ. Inf. Technol. **29**, 25143–25174 (2024). https://doi.org/10.1007/s10639-024-12836-3
15. Squire, K.D., Jan, M.: Mad city mystery: developing scientific argumentation skills with a place-based augmented reality game on handheld computers. J. Sci. Educ. Technol. **16**, 5–29 (2007). https://doi.org/10.1007/s10956-006-9037-z
16. Kamarainen, A.M., Metcalf, S., Grotzer, T., Browne, A., Mazzuca, D., Tutwiler, M.S., Dede, C.: EcoMOBILE: integrating augmented reality and probeware with environmental education field trips. Comput. Educ. **68**, 545–556 (2013). https://doi.org/10.1016/j.compedu.2013.02.018
17. Meishar-Tal, H., Ronen, M.: The impact of experiencing a mobile game on teachers' attitudes towards mobile learning. Int. J. Mobile Blended Learn. **9**(4), 21–32 (2017). https://doi.org/10.4018/IJMBL.2017100102
18. Marques, M.M., Pombo, L.: Game-based mobile learning with augmented reality: Are teachers ready to adopt it? In: Project and Design Literacy as Cornerstones of Smart Education, pp. 207–218. Springer, Singapore (2019). https://doi.org/10.4018/IJMBL.2017100102
19. Koutromanos, G., Pittara, T., Tripoulas, C.: "Clavis Aurea": an augmented reality game for the teaching of local history. Eur. J. Eng. Technol. Res. (2020). https://www.ej-eng.org/index.php/ejeng/article/view/2310
20. Costa, M.C., Manso, A., Patrício, J.: Design of a mobile augmented reality platform with game-based learning purposes. Information. **11**(3), 127 (2020). https://doi.org/10.3390/info11030127
21. Kissi, L., Dreesmann, D.: Plant visibility through mobile learning? Implementation and evaluation of an interactive flower hunt in a botanic garden. J. Biol. Educ. **52**(4), 344–363 (2018). https://doi.org/10.1080/00219266.2017.1385506
22. Creswell, J.W.: Educational Research: Planning, Conducting, and Evaluating Quantitative and Qualitative Research, 4th edn. Pearson Education Inc., Boston (2012)
23. Koutromanos, G., Jimoyiannis, A.: Augmented reality in education: exploring Greek teachers' views and perceptions. In: Reis, A., Barroso, J., Martins, P., Jimoyiannis, A., Huang, R.Y.M., Henriques, R. (eds.) Technology and Innovation in Learning, Teaching and Education, vol. 1720, pp. 31–42. Springer, Cham (2023). https://doi.org/10.1007/978-3-031-22918-3_3

Inclusive and Immersive Technologies in Education

Augmented Reality Affordances in Physics Education for Students with Intellectual Disabilities

Georgia Iatraki and Tassos A. Mikropoulos

Abstract Augmented Reality (AR) provides exciting experiences to improve learning outcomes for students with Intellectual Disabilities (ID) through educational interventions that promote inclusion. This empirical study presents a framework for designing AR experiences tailored to the unique characteristics and needs of students with ID. Recognizing the potential of AR to provide engaging and immersive learning environments, the framework aligns the strengths and preferences of students with ID with the affordances of AR to foster inclusion and progress in science classroom. An AR learning environment based on the "Model of Immersive Learning Environment for Students with Disabilities (MILES-D)" was developed, enabling students to explore abstract and invisible concepts, like the structure of matter. The study investigated both technological and learning affordances to help students with disabilities overcome learning barriers and reach their full potential. Results showed positive results in students' academic and communicative skills. The contribution of the MILES-D is emphasized, as it integrates technological affordances with the principles of the Universal Design for Learning to create inclusive learning experiences for students with ID. The study concludes by advocating for continued research and interdisciplinary collaboration to advance the introduction of AR technologies in education for students with ID.

Keywords Augmented reality · Intellectual disabilities · Inclusion · Physics

G. Iatraki (✉) · T. A. Mikropoulos
University of Ioannina, Ioannina, Greece
e-mail: g.iatraki@uoi.gr

T. A. Mikropoulos
e-mail: amikrop@uoi.gr

I. Kazanidis and A. Tsinakos (eds.), *Research on E-Learning and ICT in Education*,
https://doi.org/10.1007/978-3-031-99961-1_12

1 Introduction

Digital technology contributes to the development of autonomy, social interaction, and academic skills of students with disabilities through proper learning environments that promote motivation, interest, and engagement [1, 2]. Digital tools and applications support teaching and learning in a more efficient, personalized, and effective way, offering the opportunity to combine instructional approaches and introduce new, interactive, and adaptive options for guidance in classroom [3]. Researchers focus on immersive technologies and students prefer to use them enhancing their learning outcomes in cognitive, emotional, and affective domains [4, 5].

Affordances are the unique properties of a certain technology or environment that determine the possible actions a user can enact [6]. Concerning learning technologies, affordances lead to learning affordances, which indicate categories of learning activities which can be conducted with a certain technology [7].

AR learning affordances include collaboration, three-dimensional perception of content, visualization of the invisible, immersion and presence [8]. AR seems to improve learning experiences, as it allows students to interact with digital objects that are integrated into the natural environment [9].

Students with Intellectual Disabilities (ID) often struggle with academic, social, and practical tasks, which negatively affect their academic achievement in the classroom, as well as later in society. These deficits in intellectual functioning include "reasoning, problem-solving, organization, abstract thinking, critical judgment, and learning speed" [10]. To address these difficulties, which typically emerge in early childhood and meet the individualized needs of students, literature recommends evidence-based instructional techniques and the integration of digital technology in the learning process [11]. Adapting science content supported by digital technology could minimize students' limitations in understanding and help them acquire scientific literacy meeting the high expectations of the Framework for K-12 Science Education and Next Generation Science Standards [12–14]. In addition, research indicates that most of the scientific content taught to this population primarily focuses on Biology or Geography, highlighting a lack of attention to Physics and Chemistry [15, 16]. Physics poses greater challenges for students with ID because of its abstract concepts and the complexity of physical phenomena involved [17].

Augmented Reality (AR) has the potential to create more intuitive immersive learning environments, enhancing the understanding of complex physical phenomena. In the context of Chemistry and Physics education, where the structure of matter is a key topic, the use of models, simulations, and visualizations has been shown to improve learning outcomes [18, 19]. Additionally, studies support the integration of AR for studying matter and advancing Chemistry education [20]. However, there is a gap in literature regarding the application of AR for students with ID. The only relevant study we found is by McMahon et al. [21], which explored the use of marker-based mobile AR to teach cells and human organs to students with ID and autism spectrum disorder. Their approach, combining systematic instruction with multimedia content delivered through the AR system, led to positive learning outcomes.

AR affordances allow access to locations, increased engagement, sustained attention and collaboration [8, 9, 19]. In addition, social interaction within learning environments rich in sensory stimuli is encouraged through face-to-face communication aligned with the real-world situations. Immersion in augmented environments help students consider that they participate in a full realistic experience [22]. Regarding Physics education, AR supports more effective instructional methods such as inquiry-based learning grounded on constructivism. Learning activities based on inquiry reinforce related skills such asking questions, planning and carrying out investigations, analyzing and interpreting data, constructing explanations, and obtaining, evaluating, and communicating information [23, 24].

Based on the above, the development of an educational framework that proposes design principles for the creation on immersive learning environments addressed to students with ID arises.

Considering this lack of guidelines for the design of immersive learning environments and learning activities for students with disabilities, the "Model for the design of Immersive Learning Enactments for Students with disabilities, (MILES-D)" aligns technological affordances and learning affordances of immersive technologies with "Universal Design for Learning" (UDL) principles in order to provide students with ID with more inclusive experiences and overcome challenges during their learning process (Table 1). This model is a theoretical research-based model that incorporates the technological and learning affordances of immersive technologies, as well as the UDL principles. Delimitros and colleagues [25] validated MILES-D with students with ID focusing on their specific characteristics (limitation in learning pace, challenges in academic, cognitive, and social domains).

2 Methods

2.1 Purpose

The purpose of the current study was to investigate the impact of the proposed MILES-D that guided the design of an AR-based intervention in teaching the three states of water at a microscopic level involving students with ID (Table 1). The following research questions were formulated:

RQ1: Did the immersive AR learning environment contribute to learning outcomes of students with mild ID?

RQ2: What was the students' experience of the immersive AR intervention?

Table 1 The Model for the Design of Immersive Learning Enactments for Students with Disabilities (MILES-D)

Technological affordances of immersive technologies	
• Multisensory intuitive and real interaction • Immersion (concerning the digital content in AR) • Presence (concerns the enriched physical environment in AR) • Autonomy • Natural semantics for the representation of objects and events inside the virtual environments and worlds/the physical world is augmented by digital information in registration with physical world • User's presentation through avatars (physical representation of the user in the real environment in AR) • First person user point of view • First order experiences • Size in space and time for digital objects and events • Transduction of digital objects and events • Reification for digital objects and events	
Lead to	
Learning affordances	Udl
• Free navigation • Creation • Modeling and simulation • Multichannel communication • Collaboration and cooperation • Content presentation and/or delivery	• Engagement • Representation • Action and expression
Lead to	
Afforded learning tasks in immersive environments for students with D	
• Meaningful virtual field trips and tours • Design of safe immersive learning environments • Data presentation and interpretation, modeling, and simulation of real systems, phenomena, situations • Discussion, chatting, lectures • Meetings, role-play, social interaction. • Shared interactive tools for personal expression, presentations, and exhibitions	

2.2 The Immersive AR Learning Environment

For the design of the academic content for students with ID we followed MILES-D and developed a molecular dynamic simulation for the three states of water. The validation of the scientific terminology and the application questions were conducted with experts on science education who participated in a pilot study [25, 26], combined with the relevant literature and the curricula of primary and secondary education regarding the structure of matter [14, 18]. Specifically, we considered the main cognitive characteristics of mild ID and followed the guidelines for typically developing children in fifth grade level (10–11 years old). The classification of ID indicates that

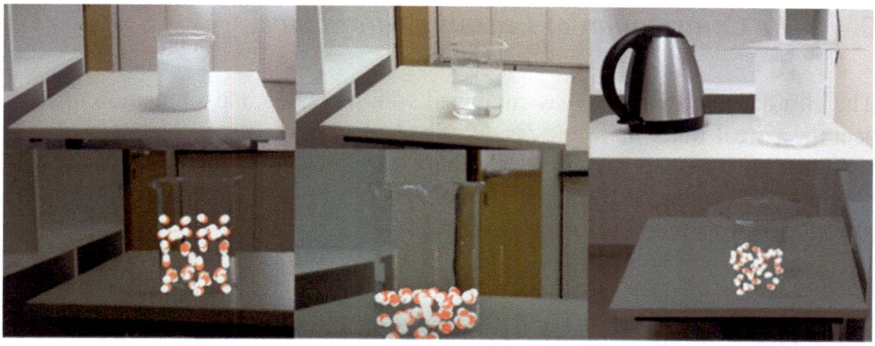

Fig. 1 Real macroscopic view (up) and augmented microscopic view (down) for the three states of water (left to right: solid, liquid, gas)

the IQ range of 50–69 is equivalent to the mental age of 9–12 years of typically developing children [27]. In those ages students are taught the structure of matter (atoms and molecules), the particle model and the three states of matter [28]. The immersive AR learning environment included the real world of a classroom, where a real beaker was placed, augmented with virtual water molecules in the three states (solid, liquid, gas) (Fig. 1). We used the Magic Leap One™ AR device, which allowed the direct observation of the physical world through semitransparent AR glasses, augmented with the digital content.

2.3 The Contribution of AR Technological Affordances

The spatial and temporal scales, transduction, and reification were used for the 3D representation of the molecules. The representation followed the atomic structure characteristics, namely the van der Waals radii [29] of oxygen (1.4 Å) and hydrogen (1.2 Å), their distance between their centers (0.957 Å), and their orientation angle (104.5°). The virtual molecules were arranged and moved depending on their state (solid, liquid gas). The water molecule was represented as a two-color model, based on the Corey-Pauling-Koltun coloring scheme convention [30].

The affordances of immersion and presence applied to align the virtual objects with the real world, providing real-time interactions. Regarding autonomy, virtual molecules act independently following the Physics laws. Transduction bridges the gap between the physical world (macroscopic view) and the virtual one (microscopic view) allowing users to see the unseen. Reification represents the arrangement and movement of molecules in their three different states. Presence was enhanced when gaseous molecules moved freely in the real classroom.

Each student could move freely in the classroom wearing the AR glasses, having first person user point of view and first order experiences.

2.4 The Contribution of Learning Affordances

The affordance of creation was combined with modeling and simulation to represent the virtual water molecules as described above and present the proper content presentation that integrated in the learning environment.

The users navigated freely in the real room to observe and study the physical phenomena from their own point of view.

2.5 The Contribution of UDL

Universal Design for Learning is an educational framework based on cognitive neuroscience research that aims to provide a flexible and accessible educational environment to all diverse learners, including students with disabilities [31, 32]. The UDL framework is organized through three main principles: engagement, representation, action, and expression [33] that could effectively eliminate barriers in the educational environment.

The parallel experience of both the macroscopic and microscopic worlds in the real environment aimed at the participants engagement with the learning environment.

The presentation of the teaching content using appropriate terminology together with the special education instructional techniques utilized aimed at participants' perception and comprehension.

The participants free navigation, observation, and interaction with the AR environment contributed to their physical action, expression and communication, and executive functions.

2.6 Afforded Learning Tasks in Immersive Environments for Students with ID

Regarding the proposed learning tasks that MILES-D describes, the five following dimensions were applied:

- Meaningful virtual field tour.
- Design of a safe immersive learning environment.
- Data presentation and interpretation, modeling, and simulation of a real system, phenomena, situations.
- Discussion, chatting, lectures.
- Meetings, role-play, social interaction.

The real macroscopic situation as well as the modeled and simulated microscopic views consisted of a safe immersive learning environment, where students with ID could study the structure of matter and handle possible harmful situations (e.g., boiled water, and water vapor).

The integration of inquiry-based learning, where students explore situations, allowing them to construct knowledge and develop inquiry skills in the immersive AR environment, aimed at the development of discussions and interpretation of the physical phenomena.

2.7 Experimental Design, Participants, Setting, and Procedures

A single-subject AB research design was applied to evaluate the AR learning environment. AB design includes repeated measurements of learning outcomes across several phases, baseline (A) and main intervention (B) in this case. These phases allow correlational conclusions to be drawn between the corresponding performances [34].

Six students with mild ID participated in the AR intervention for whom parental consent was obtained. Two male and four female Greek students ages 14–17 years were met the following inclusion criteria: (a) a diagnosis of mild ID, (b) adequate visual and auditory discrimination, (c) ability to respond verbally, (d) basic reading and writing skills, and (e) basic computer skills. The main difficulty of the students was reading comprehension, especially when students had to use Physics terms and interpretations of physical phenomena. The students also required prompting and oral instructions to complete the educational activities. The sessions took place in a one-to-one setting in the students' school.

Baseline data were collected over 13 sessions to assess previous knowledge of the six students regarding the states of water. The intervention included six sessions and began simultaneously for the six students. Each session lasted 20 to 30 minutes depending on the student's progress. Three students received Physics content through systematic instruction and the other three participated in an inquiry process answering open-ended questions. Both instructional models were supported by the techniques of task analysis, prompting, and time delay.

In each session, students answered seven questions classified to the first three levels of the revised Bloom taxonomy. In addition, we collected maintenance data six times, following the criterion of completing the three main topics: describing water molecules, their arrangement and movement in each state. Two weeks after the intervention, a generalization phase was conducted, where structured questions aimed to expand students' acquired knowledge. The students observed a different molecule that coexisted with water molecules in the real classroom and were asked to identify the "unknown molecule", describe (shape, color, size) and compare it with water molecules. This session was designed to assess inquiry skills in line with the eight practices proposed by the National Research Council [14].

Finally, we collected social validity data on students' experience related to (a) students' learning outcomes, (b) content quality, the ease of using the AR glasses and the effectiveness of the molecule design in understanding, (c) levels of engagement and motivation, (d) sense of presence, and (e) enjoyment of participating in the study. This instrument was based on an adapted 3-point Likert questionnaire—disagree, neither disagree nor agree, agree—of 18 items based on the Learning Object Evaluation Survey for Students [35].

3 Results and Discussion

3.1 Learning Outcomes and Social Validity Outcomes

According to the first research question, Fig. 2 shows the number of correct responses during the AR intervention given by students under each instructional model (inquiry on the left, systematic instruction on the right).

Visual analysis involved both qualitative and quantitative approaches, which were presented on individualized graphs of the participants based on the several phases of the single-subject design [36]. During baseline, the six students presented very low levels of correct answers indicating a decelerating trend (focusing the two last baseline sessions where a comparison was conducted). Entering intervention, the students increased their correct responses, recording an important change in level. Visual analysis of the individualized graphs showed a functional relationship between AR-based

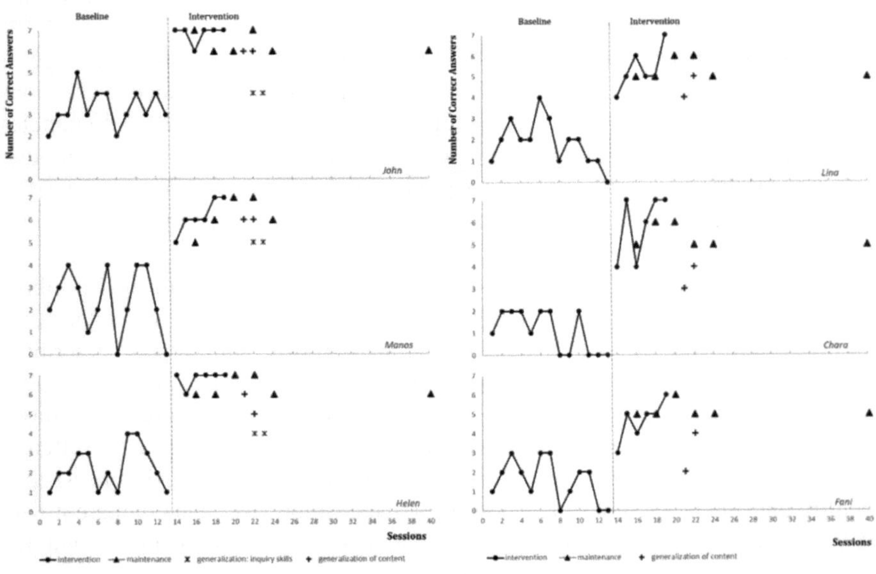

Fig. 2 Number of correct responses for students in AR-based intervention. Baseline: 0–13, intervention: 14–20, maintenance: 16–24, generalization: 22–23

interventions and the correct responses on session probes for the six students. The maintenance data were high, too.

Regarding the second research question, the six students assessed the AR intervention by responding to 18 social validity questions about their experience (Table 2). The results showed positive feedback on learning, quality, engagement, and presence. Most responses (13 out of 16) received a rating of 3 (agree), reflecting strong approval. The six students agreed that the AR glasses were easy to use and improved their learning experience. They enjoyed using the AR glasses and felt more engaged with the content, noting a strong sense of presence, as if the molecules were in their physical space. The students were highly satisfied and expressed interest in using the AR glasses for other topics.

3.2 The Contribution of MILES-D

The positive learning outcomes and strong social validity suggest that the MILES-D components like the free navigation in a safe environment enhanced by 'invisible to the naked eye' molecules, fostered effectively students' inquiry and conceptual understanding skills, areas where they often face challenges. Students were able to ask questions (e.g., how can I see what ice is made of? or how many phases water have?), analyze and interpret data (e.g., gas molecules move very quickly, while in the solid state remain in fixed positions and oscillate), evaluate and communicate information (e.g., the new unknown object is a molecule. It moves faster than vapor molecules). Additionally, it seems that the multiple means of representation (the macroscopic and microscopic views), the engagement through the AR glasses and action-expression helped students reach their full learning potential.

The high spatial presence scores appeared to enhance students' engagement and interest in the learning process, aligning with findings by Dalgarno and Lee [37] for virtual environments and Carreon et al. [38] for students with disabilities. Despite challenges with distraction and lack of concentration, students felt present in the augmented world, particularly when they removed the real cap from the beaker to release the water vapor molecules into the room.

Integrating digital content in the physical world supported both the learning process and students' understanding [39]. The visualization of invisible molecules bridged the gap between the virtual and real worlds, transforming the learning of abstract concepts typically presented in textbooks and other media. This immersive approach promoted conceptual understanding by enabling a comprehensive perception of abstract content in real time [5]. The scientifically accurate representation of molecules enhanced students' discriminative ability, supporting cognitive processes and reducing mnemonic load. This was especially beneficial for students with ID, contributing significantly to improved outcomes. Students also recognized the consistent structure of molecules across phases and identified the empty space between them, using terms like 'empty space' or 'nothing' to describe their observations.

Table 2 Social validity data (3-point Likert scale)

Questions		Mean ± SD
A. Learning		
1.	Did the lesson with the glasses help you learn?	3 ± 0
2.	Did the molecules, as you saw them, help you learn?	3 ± 0
B. Quality		
3.	Did the molecules, as you saw them through the glasses, assist you?	3 ± 0
4.	Did you understand easily how to use the glasses?	3 ± 0
5.	Were the glasses easy to use?	3 ± 0
6.	Were the molecules well designed?	3 ± 0
C. Engagement		
7.	Did you like what you saw through the glasses?	3 ± 0
8.	Did the glasses encourage you to learn?	3 ± 0
9.	Would you like to use the glasses again?	3 ± 0
10.	Would you like to use the glasses in other subjects as well?	3 ± 0
D. Presence		
11.	Did the molecules appear to be in the same space where you were?	3 ± 0
12.	Did you want to approach and touch the molecules you saw through the glasses?	2,6 ± 0.5
13.	Did you consider avoiding any molecules?	2,3 ± 0.5
14.	Did you feel like you were among the molecules you saw?	3 ± 0
15.	Did you see the molecules as if you were looking through a screen?	1 ± 0
16.	Did it seem as if you were observing the molecules being in the classroom where you were?	3 ± 0
E. Open ended questions		
17.	What did you like the most?	
"I liked seeing molecules so close!" (John) "I liked the most when the molecules went everywhere!" (Manos, Chara) "I liked everything!" (Helen) "I liked the moment I moved the cap, and the molecules went away!" (Lina) "It was perfect! I liked seeing the unseen molecules!" (Fani)		
18.	What did you not like at all?	
I liked everything.		

Our findings align with existing research, emphasizing the significance of sensory and visual contexts in enhancing motivation and engagement during the intervention. Students perceived the augmented environment as a part of their reality [40]. These results are consistent with McMahon and colleagues' [21], who showed that students with ID using portable AR devices maintained focus and exhibited high levels of participation in Biology lessons. Immersive AR provided a student-centered approach, connecting learning to the individualized profile of each student [41]. This approach supports learning in both general and special education settings [42].

4 Conclusions

This study highlighted the benefits of an immersive AR intervention for promoting the inclusion of students with ID in Physics education. These benefits were achieved by leveraging AR affordances and adhering to the principles of universal design as outlined in the MILES-D framework.

This work contributed to the field by demonstrating that AR supports the acquisition of academic content by students with ID, particularly in learning terminology and understanding concepts within the general curriculum. It also emphasized the importance of leveraging AR's benefits in teaching abstract or invisible phenomena [43]. Additionally, it contributed significantly to removing students' barriers regarding cognitive and mnemonic processes, inquiry skills, and social skills [44]. By aligning design processes and student profiles, AR interventions can create meaningful opportunities to ensure the development of personalized and preferred AR experiences. We believe that the AR intervention, with its innovative design and positive impact on student engagement and motivation, represents a promising potential step toward the future of Physics inclusion that engages all students in teaching and learning of abstract content.

Despite the potential of immersive AR for students with ID, challenges remain in addressing the diverse needs associated with varying levels of ID. By identifying and leveraging students' strengths such as attention to detail (discrimination process), the ability to block out distractions (concentration), and responsiveness to guidance (following instructions), research can offer concrete and actionable recommendations for AR design. Design frameworks like MILES-D can play a crucial role in standardizing approaches to accessibility, usability, and the development of robust universal design principles for AR experiences. Such frameworks help ensure that AR technology is both inclusive and effective for all learners.

References

1. Barlott, T., Aplin, T., Catchpole, E., Kranz, R., Le Goullon, D., Toivanen, A., Hutchens, S.: Connectedness and ICT: opening the door to possibilities for people with intellectual disabilities. J. Intellect. Disabil. **24**(4), 503–521 (2020)
2. Kellems, R.O., Cacciatore, G., Osborne, K.: Using an augmented reality–based teaching strategy to teach mathematics to secondary students with disabilities. Career Dev. Transit. Except. Individ. **42**(4), 253–258 (2019)
3. de Jong, T.: Moving towards engaged learning in STEM domains; there is no simple answer, but clearly a road ahead. J. Comput. Assist. Learn. **35**(2), 153–167 (2019)
4. Akçayır, M., Akçayır, G.: Advantages and challenges associated with augmented reality for education: a systematic review of the literature. Educ. Res. Rev. **20**, 1–11 (2017)
5. Wu, H.-K., Lee, W.-Y., Chang, H.-Y., Liang, J.-C.: Current status, opportunities and challenges of augmented reality in education. Comput. Educ. **62**, 41–49 (2013)
6. Norman, D.A.: The Design of Everyday Things. Basic Books, New York (2013)
7. Mantziou, O., Papachristos, N.M., Mikropoulos, T.A.: Learning activities as enactments of learning affordances in MUVEs: a review-based classification. Educ. Inf. Technol. **23**(4), 1737–1765 (2018)
8. Arena, F., Collotta, M., Pau, G., Termine, F.: An overview of augmented reality. Computers. **11**(2), 28 (2022)
9. Chen, Y.-C., Chi, H.-L., Hung, W.-H., Kang, S.-C.: Use of tangible and augmented reality models in engineering graphics courses. J. Prof. Issues Eng. Educ. Pract. **137**(4), 267–276 (2011)
10. American Psychiatric Association (ed.): Diagnostic and Statistical Manual of Mental Disorders, 5th edn. American Psychiatric Association, Washington (2013). https://doi.org/10.1176/appi.books.9780890425596
11. Mikropoulos, T., Iatraki, G.: Digital technology supports science education for students with disabilities: a systematic review. Educ. Inf. Technol. **28**, 3911–3935 (2022)
12. Ainscow, M., Booth, T., Dyson, A., Farrell, P., Frankham, J., Gallannaugh, F., et al.: Improving Schools, Developing Inclusion. Routledge, London (2006)
13. Individuals with Disabilities Education Improvement Act (IDEA), 20 U.S.C. §1400, H.R. 1350 (2004)
14. National Research Council: A Framework for K-12 Science Education: Practices, Crosscutting Concepts, and Core Ideas. The National Academies Press, Washington (2012)
15. Knight, V., Creech-Galloway, C., Karl, J., Collins, B.: Evaluating supported eText to teach science to high school students with moderate intellectual disability. Focus Autism Other Dev. Disab. (2017). https://doi.org/10.1177/1088357617696273
16. Miller, B., Krockover, G., Doughty, T.: Using iPads to teach inquiry science to students with a moderate to severe intellectual disability: a pilot study. J. Res. Sci. Teach. **50**(8), 887–911 (2013)
17. Terrazas-Arellanes, F., Gallard, M., Strycker, A.L., Walden, E.: Impact of interactive online units on learning science among students with learning disabilities and English learners. Int. J. Sci. Educ. **40**(5), 498–518 (2018)
18. Bellou, I., Papachristos, N.M., Mikropoulos, T.A.: Digital learning technologies in chemistry education: a review. In: Sampson, D., Ifenthaler, D., Spector, J., Isaías, P. (eds.) Digital Technologies: Sustainable Innovations for Improving Teaching and Learning, pp. 57–80. Springer, Cham (2018)
19. Dargan, S., Bansal, S., Kumar, M., Mittal, A., Kumar, K.: Augmented reality: a comprehensive review. Arch. Comput. Methods Eng. **30**(2), 1057–1080 (2023)
20. Jiménez, Z.A.: Teaching and learning chemistry via augmented and immersive virtual reality. ACS Symp. Ser. **1318**, 31–52 (2019)
21. McMahon, D., Cihak, D., Wright, R., Bell, S.: Augmented reality for teaching science vocabulary to postsecondary education students with intellectual disabilities and autism. JRTE. **48**, 38–56 (2016)

22. Dede, C.: Immersive interfaces for engagement and learning. Science. **323**, 66–69 (2009)
23. Burns, J.C., Okey, J.R., Wise, K.C.: Development of an integrated process skill test: TIPS II. J. Res. Sci. Teach. **22**(2), 169–177 (1985)
24. Lou, Y., Blanchard, P., Kennedy, E.: Development and validation of a science inquiry skills assessment. J. Geosci. Educ. **63**(1), 73–85 (2015)
25. Delimitros, M., Stergiouli, A., Iatraki, G., Koutromanos, G., Mikropoulos, T.: A model for the design of immersive learning enactments for students with intellectual disability. In: 10th 2022 International Conference of Software Development and Technologies for Enhancing Accessibility and Fighting Info-Exclusion (DSAI 2022), pp. 1–7. ACM, New York (2022)
26. Iatraki, G., Delimitros, M., Vrellis, I., Mikropoulos, T.: Augmented and virtual environments for students with intellectual disability: design issues in science education. In: 2021 International Conference on Advanced Learning Technologies (ICALT), Tartu, Estonia (online), pp. 381–385 (2021)
27. Bhaumik, S., Kiani, R., Michael, D.M., Gangavati, S., Khan, S., Torales, J., Ventriglio, A.: Intellectual disability and mental health: an overview. Int. J. Cult. Ment. Health. **9**(4), 417–429 (2016)
28. Johnson, P.: Progression in children's understanding of a "basic" particle theory: a longitudinal study. Int. J. Sci. Educ. **20**(4), 393–412 (1998)
29. Batsanov, S.S.: Van der Waals radii of elements. Inorg. Mater. **37**, 871–885 (2001)
30. Koltun, W.L.: Precision space-filling atomic models. Biopolymers. **3**(6), 665–679 (1965)
31. Almeqdad, Q.I., Alodat, A.M., Alquraan, M.F., Mohaidat, M.A., Al Makhzoomy, A.K.: The effectiveness of universal design for learning: a systematic review of the literature and meta-analysis. Cogent Educ. **10**(1) (2023). https://doi.org/10.1080/2331186X.2023.2218191
32. Meyer, A., Rose, D.H., Gordon, D.: Universal Design for Learning: Theory and Practice. CAST Professional Publishing, Chicago (2014)
33. CAST: Universal Design for Learning Guidelines version 3.0. https://udlguidelines.cast.org (2024)
34. Horner, R.H., Carr, E.G., Halle, J., McGee, G., Odom, S., Wolery, M.: The use of single-subject research to identify evidence-based practice in special education. Except. Child. **71**(2), 165–179 (2005)
35. Kay, R.H., Knaack, L.: Developing learning objects for secondary school students: a multi-component model. Interdiscip. J. Knowl. Learn. Objects. **1**, 229–254 (2005)
36. Iatraki, G., Mikropoulos, T.A.: Augmented reality in physics education: students with intellectual disabilities inquire the structure of matter. PRESENCE. **31**, 89–106 (2023)
37. Dalgarno, B., Lee, M.J.W.: What are the learning affordances of 3-D virtual environments? Br. J. Educ. Technol. **41**(1), 10–32 (2010)
38. Carreon, A., Smith, S.J., Mosher, M., Rao, K., Rowland, A.: A review of virtual reality intervention research for students with disabilities in K-12 settings. J. Spec. Educ. Technol. **37**(1), 82–99 (2020)
39. Dunleavy, M., Dede, C., Mitchell, R.: Affordances and limitations of immersive participatory augmented reality simulations for teaching and learning. J. Sci. Educ. Technol. **18**(1), 7–22 (2009)
40. Radu, I., Schneider, B.: What can we learn from augmented reality (AR)?: benefits and drawbacks of AR for inquiry-based learning of physics. In: CHI Conference on Human Factors in Computing Systems (CHI '19), pp. 1–13. ACM, New York (2019)
41. Munnerley, D., Bacon, M., Wilson, A.G., Steele, J., Hedberg, J., Fitzgerald, R.N.: Confronting an augmented reality. Res. Learn. Technol. **20**, 19189 (2012)
42. Arici, F., Yildirim, P., Caliklar, Ş., Yilmaz, R.M.: Research trends in the use of augmented reality in science education: content and bibliometric mapping analysis. Comput. Educ. **142**, 1–23 (2019)

43. Baragash, R.S., Al-Samarraie, H., Alzahrani, A.I., Alfarraj, O.: Augmented reality in special education: a meta-analysis of single subject design studies. Eur. J. Spec. Needs Educ. **35**(3), 382–397 (2020)
44. Cheng, K.H., Tsai, C.C.: Affordances of augmented reality in science learning: suggestions for future research. J. Sci. Educ. Technol. **22**, 449–462 (2013)

Enhancing Participation of Students with Autism in Inclusive Educational Robotics: An Action Research

Evaggelia Tsiomi, Artemis Maistrelli, Andromachi Nanou, and Dimitris Karampatzakis

Abstract Educational Robotics (ER) provides opportunities for students with autism to leverage their strengths and be included in peer groups. As ER becomes an integral part of the curriculum, the urgent need to develop specific strategies for their inclusion becomes increasingly apparent. To address this need, an action research study was conducted in non-typical educational settings to evaluate the effectiveness of the Search and Share Strategy (SaSS), a specific strategy designed to enhance the participation of students with autism in assembling and block programming when they take part in Inclusive Educational Robotics activities (IERa). Six students diagnosed with autism received remote training in SaSS. After receiving this training, they participated in face-to-face IERa alongside twelve neurotypical peers, in groups of three, using WeDo 2.0 and Spike LEGO Education Robots (ERs). This study examined the transferability and impact of the SaSS training across both instructional contexts. The findings indicated that the remote SaSS training had a positive impact on the participation of students with autism in IERa fostering positive attitudes among neurotypical peers toward the inclusion of students with autism.

Keywords Educational robotics · Autism · Inclusion · Participation · Block programming

E. Tsiomi · A. Nanou · D. Karampatzakis
Department of Informatics, Democritus University of Thrace, Kavala, Greece
e-mail: ananou@cs.duth.gr

D. Karampatzakis
e-mail: dkara@cs.duth.gr

E. Tsiomi (✉) · A. Maistrelli · A. Nanou
Interdisciplinary Network for Special and Intercultural Education "Include", Thessaloniki, Greece
e-mail: etsiomi@cs.duth.gr

A. Maistrelli
e-mail: artemis@include.edu.gr

I. Kazanidis and A. Tsinakos (eds.), *Research on E-Learning and ICT in Education*,
https://doi.org/10.1007/978-3-031-99961-1_13

1 Introduction

Educational Robotics (ER) has become a vital educational activity offering a dynamic platform that introduces students to computational thinking, creativity, and collaborative behavior [1]. ER serves as a flexible tool so that teachers can adapt to the educational needs of their students and course requirements.

Students with Autism Spectrum Disorder (ASD) often struggle with understanding basic social rules and experience anxiety during social interactions, leading to reduced participation in social activities compared to their peers [2, 3]. ASD can be manifested at three Levels: Level 1, for moderate social and adaptability difficulties, Level 2 for significant social and adaptability challenges, and Level 3 for severe social and adaptability issues.

Although students with ASD find it difficult to participate effectively in social situations and collaborate with peers, in ER activities they are offered opportunities to be included [4, 5]. In Inclusive ER activities (IERa) research shows that the implementation of specific strategies fosters collaboration in assembling, programming, and problem-solving [5–7]. Research in Inclusive ER environments shows enhanced social and cognitive skills in students with ASD [8–10]. ER had been recommended as a suitable inclusive activity that provides opportunities for students with ASD to capitalize on their strengths and be included in peer groups [11]. Positive findings about participation, collaboration, and improvement of the social and cognitive skills of students with ASD have been confirmed [6, 9, 12].

Important parameters that enhance the inclusion of students with ASD in ER activities are: (a) the structuring of the environment, (b) the structuring of the activities, (c) the structuring of the roles during collaboration, and (d) the structuring of the process of problem solving in the context of achieving short-term and long-term goals [9]. Designing learning contexts to meet the diverse needs of all students, combined with targeted strategies, enhances the participation of students with ASD in assembling and programming ER activities [13]. Structured scenarios and defined team roles promote the participation [7, 14]. Distinct roles include "Architect" (plans and manages coding), "Supplier" (provides parts and coding blocks), and "Builder" (assembles and organizes). Search and Share Strategy (SaSS) designed for students with ASD to succeed as "Suppliers" in IERa and their interaction with peers [9, 15]. SaSS is strategically designed to enhance the abilities of individuals with ASD in blocked-base programming, enabling them to effectively serve as "Suppliers" in collaboration with the "Architect" and "Builder." Students are taught how to follow the instructions designed in a manual and implement the five steps of SaSS to provide bricks to the "Builder" and thus become effective "Suppliers" and participate in interactions with all members of IER team [9].

The present study investigates the perspectives of participation of students with ASD in IERa after receiving remote SaSS training.

2 Method and Materials

To achieve our goals, we applied action research according to Lewin's model [14] through four cycles: Plan, Act, Observe, and Reflect. This methodology was chosen for its effectiveness in real-world settings, enhancing problem-solving skills and facilitating collaborative improvement through feedback and evaluation [8]. A control group was not selected as the action research we are pursuing is interested in recording the changes observed during the research process and focuses on an in-depth understanding of the phenomena in the specific context, making the existence of a control group less necessary [16].

2.1 *Research Questions*

The action research aimed to evaluate the influence of the SaSS remote training on enhancing the participation of students with ASD in assembling and block programming in IERa. To address this objective, the study seeks to answer the following research questions:

1. RQ1. How does SaSS remote training contribute to the participation of students with ASD in LEGO assembly activities?
2. RQ2. How does SaSS enhance the participation of students with ASD in assembling and block programming during face-to-face IERa?
3. RQ3. What is the correlation between students' participation in remote SaSS training and their participation in assembly and block programming during face-to-face IERa? Can participation in remote training predict participation in face-to-face IERa?
4. RQ4. What are the views of typical peers and caregivers on students with ASD participation in IERa?

2.2 *Participants*

A total of eighteen (18) students participated in the action research, consisting of six (6) with ASD and (12) twelve neurotypical peers. The age range of the participants was from 12 to 30 years. The sessions were implemented using convenience sampling, where researcher had access due to social acquaintances. The age heterogeneity of the sample was imposed by the non-typical education context in which action research was conducted. Heterogeneity in this sense makes the research results generalizable and more widely applicable to similar settings [16]. In addition, the action research and educational process involved three team coaches and one observer, all of whom were women. Their ages ranged from 20 to 45 years old, and they were experienced special educators.

2.3 Place, Schedule, and Materials

The action research was implemented in two (2) Stages:

1. Remote (Stage I): individuals collaborated with caregiver support, using the LEGO's Kit Education Simple Machines Set.
2. Face-to-Face (Stage II): involved group work with two neurotypical peers and a student within the ASD spectrum, using LEGO Education WeDo 2.0 and Spike tools.

Sessions were strategically planned and teams used tablets with LEGO educational infrastructure and printed coding blocks. Each Stage is implemented in three sessions.

3 Educational Design

The educational design process adhered to the ADDIE Model.

Analysis. In the initial phase, an extensive needs assessment was carried out to identify the educational requirements of both students with ASD and their caregivers, as well as the needs of neurotypical peers.

Design. The educational methodology and specific goals were subsequently articulated. Informed by prior research demonstrating the efficacy of the SaSS training in enhancing the involvement of students with autism in IERa [9], the intervention was carefully designed. The overarching objective was to empower participants with ASD to autonomously fulfill the role of suppliers within teamwork scenarios alongside their peers in diverse capacities, including architects and builders.

In this critical phase, training materials were systematically developed to facilitate the implementation of SaSS training. To structure the social framework for implementation, the social scripting method was developed. This method involves the incorporation of authentic images depicting social situations that explode from the strong frequency of memory recall observed in students with ASD for social scenarios. The scenarios were presented to the mixed group in the face-to-face sessions. Additionally, to facilitate block-based programming, printed and multi-layer printed coding blocks were special devices. These cards were adapted to guide participants in arranging coding blocks in a precise order, increasing their improvement in block-based coding activities.

Implementation. Remote *Sessions (Stage I)*. Students with ASD met with their caregiver and coach via a tablet or PC from their room and introduced themselves. The two participants sat at a desk with the materials in front of them. Caregivers positioned the camera so the observer could see both the materials and the child-caregiver interaction. These three sessions focused solely on assembling using LEGO equipment. The first cycle of action research, included all four phases, culminating in the revised design of the educational intervention. Subsequently, the second cycle was

Fig. 1 First remote session

iterated, involving a redefinition of certain elements within the educational design. The revised plan was implemented in two sessions.

Face-to-Face Sessions (Stage II). At Stage II, each of the three students (one with ASD and two neurotypical peers) followed the same Strategy but used different training materials. The child with ASD applied the Strategy taught during Stage I as a supplier. He/she searched for the necessary LEGO piece indicated by the manual and shown by the neurotypical student. Once found, the piece was given to the other neurotypical student to continue assembling. In the second cycle of the action research, the educational materials were adapted, and techniques such as reinforcement of desired behaviors by the coaches were employed.

Equipment. *Remote sessions (Stage I).* Simple machines selected for assembling by the kit LEGO education simple machine set. During the remote sessions, the student builds the following machines (Fig. 1).

Face-to-face sessions (Stage II). During the face-to-face sessions, the students build the following machines with LEGO Education WeDo 2.0 and Spike tools (Fig. 2).

The students collaboratively constructed satellites using the provided kit and subscribed to program each functionality. Despite the LEGO Education WeDo 2.0 limitations in variety and programming complexity, it successfully installed a basic understanding of block programming. The kit offered two primary options for robot movement: circular motion, as demonstrated in the initial scenario, and forward-backward motion. Milo, an endearing robot, exhibited the capability to navigate both forward and backward at variable speeds. For the third segment, we transitioned to a more advanced level of block programming by upgrading to the Spike kit. The heightened difficulty level of the Spike kit emanates from the inclusion of various motors and sensors, amplifying the complexity of the programming tasks.

The array of choices in both construction and programming rendered the process simultaneously challenging and intriguing. Commencing the lesson, we posed the question of how incorporating a second motor would impact the robot's movement.

Fig. 2 Third face to face session

Finally, data collected from the educational intervention were analyzed and presented.

3.1 Data Collection

A mixed methodology for data collection and analysis ensures research validity and reliability. Various tools both quantitative and qualitative were utilized to collect data within a specified time frame, offering insights into the participation of students with ASD.

3.2 Research Tools

Six Research Tools (RT) were used: Caregivers' questionnaire (RT1) aimed at understanding caregiver cooperation with student with ASD, before the SaSS remote training. Observation protocol (RT2) is a rubric to assess the level of student's autonomy during SaSS remote or face-to-face training by an independent observer. Daily diary of the child's participation during SaSS remote or face-to-face training (RT3) recorded by the participant observer, interviews with typically developed student exploring their personal IERa experiences (RT4), caregiver interviews (RT5) evaluate perceptions of what students with disabilities gained from the IERa, Likert recorded the feelings of students with ASD about how they experienced the IERa (RT6).

3.3 Data Analysis

The research employed a concurrent parallel design for data collection and analysis, ensuring reliability and validity. Triangulation of data from interviews, questionnaires, and observations was conducted simultaneously.

4 Results

4.1 Findings Related to the First RQ

How can remote training on the SaSS contribute to the participation of students with ASD in LEGO assembly activities?

Regarding First Action Research Cycle Before SaSS remote training, caregivers emphasized the importance of engaging in a shared activity with their child and stressed the significance of providing support in acquiring skills and experiences as for their participation in LEGO activities supporting their child, all caregivers were positive as depicted by the RT1 data analysis. According to caregivers' perspectives, none of the students could participate in assembly activities using a guide manual without support. Additionally, even the students in Level 1 were not able to collaborate in the assembling and programming of educational robots. Caregivers' expectations from the remote SaSS training included that the students participate creatively, interact, and acquire new skills, self-confidence, and social inclusion in peer groups.

Indeed, according to the analysis of the observation protocol data (RT2), collected by the independent observer during SaSS remote training no student was autonomous and needed guidance from the caregivers. The support was to some extent proportional to students with ASD level of functioning (ASDL1 or 2 or 3). Students A and B with a need for support (ASDL1) and student E with a need for significant support (ASDL2) were evaluated with a degree of autonomy: 2.0 and 3.0 respectively, as they participated as suppliers with verbal or occasional guidance from their caregivers. Students C, and D with significant support (ASDL2) and F with a need for very significant support (ASDL3) were assessed with Degree of autonomy: 1.0 and 2.2 as they did not participate or participate with physical guidance in the assembly process by their caregiver. As noted in the diaries by the observer and the caregivers' interviews (RT3 and RT5), students remained uninvolved and/or acted with physical guidance, did not achieve joint focus, and even if they performed some part of the process they did not continue, and their actions did not have coherence and a sequence. Caregivers express the need for the steps to be presented during the process so that the students can effectively use the SaSS during assembling. The process is repeated using SaSS and the caregivers, and the independent observer assesses the participation of the students with ASD (see Fig. 3).

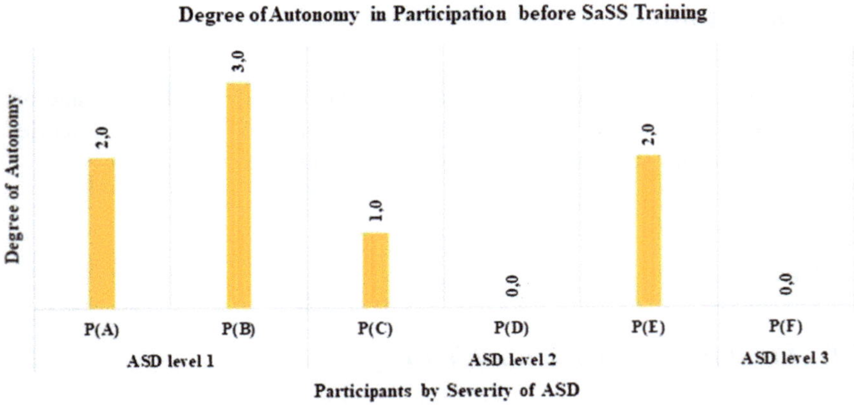

Fig. 3 Mean degree of autonomy in participation before SaSS training as assessed during first Action Circle by the independent observer

Reflecting and re-planning. In reflecting after the first action circle challenges were noted concerning (a) how the SaSS was to be more effectively presented and (b) how the equipment was to be settled for the researcher observer to have the opportunity to assess the participation of the students during the online training. As for the (a) In considering the perspective of most caregivers, their emphasis on the child's responsive participation sustained attention, and the application of specific learning strategies, such as imitation, emerges as a common theme during their collaborative efforts. In addition, the need for the effective presentation of SaSS was imperative as student (F) experienced stress upon hearing the researcher's voice by the microphone. However, challenges were noted during the online interventions conducted by the researcher using a microphone. Student F experienced stress upon hearing the researcher's voice, requiring a pause to regain composure and refocus with the caregivers' support. For the prose to become specific and visualize a social scenario on the steps of the SaSS was discussed and decided to be prepared and sent to the caregivers to be printed to be placed in front of the students after the SaSS with the support of their caregivers through the digital social scenario presented online.

As for (b) the placement of equipment and cameras as observations that were written in the researcher's diaries (RT3) prompted that there were difficulties in observing the students and how they collaborated with their caregiver's specific adjustments. The objective was to provide the observer with a comprehensive view of the participant's role as a supplier. However, a notable challenge was identified—no established communication mode between the researcher and the caregiver was in place to guide the process as needed. The caregiver played a pivotal role, turning the page of instructions and reminding the child of the next step. The child actively searched for the right piece, compared it to the picture in the instruction book, and assembled it, with the caregiver guiding physical or verbal prompts as needed. In addressing communication challenges, especially with students with ASD Levels 2 and 3, it was specified to use the chat for effective interaction between the researcher and the caregiver. The remote SaSS training process as adapted during first Stage became more autism-friendly and effective.

Regarding the Findings of the Remote Second Action Research Cycle In the session of SaSS training, two students with ASD Level 1, P(A) and P(B) were evaluated through the RT2, by the independent observer at the highest degree of autonomy (3.8 and 3.6) as suppliers with their caregivers. The observer's diary entries indicated that these students successfully applied SaSS with the support of their caregivers after he digital social scenario presented online. They demonstrated readiness to adeptly follow the steps of the SaSS with reminders of the steps by their caregivers. Conversely, the other two students (D and F) were assessed as needing moderate support (Degree 2) and requiring verbal guidance. Additionally, two students (C and E) were identified as needing occasional guidance (Degree 3).

Despite the supportive involvement of their caregivers these individuals required more assistance, including verbal guidance or reminders to assume the role of a supplier. This additional support was deemed necessary, even though the caregivers were actively engaged and supportive in the process. The observer's diary entries highlighted the effectiveness of the social scenario, a sentiment that resonated with the caregivers, as revealed in the interviews. The strategic use of the social scenario significantly aided participants, providing valuable visual guidance from the very first session. This collaborative approach, involving participants reading the social scenario with caregivers' support, proved instrumental in preparing them for the implementation of the SaSS to assume the role of suppliers.

On a positive note, the caregiver of P(D) demonstrated active support by patiently waiting for each step execution. This caregiver employed effective verbal guidance explaining the SaSS steps which proved beneficial for the student. The caregiver's involvement played a key role in the participants' successful enhancement of the SaSS.

Based on the observation protocol, a notable trend emerged in the last two sessions. All participants progressively gained independence in their roles as suppliers. Participants initially assessed at ASD Levels 1 and 2 demonstrated the ability to function autonomously, assembling using the SaSS. Those at Level 3, while requiring physical support in the first session, notably improved their participation quality after two sessions, ultimately reaching autonomy with visual support.

Students with ASD Level 1, P(A), and (B) achieved autonomous SaSS use from the first action, maintaining autonomy. Students with ASD Level 2 P(C), (D), and (E) initially utilized the strategy with visual guidance, achieving full autonomy by the last session. The severity of their ASD influenced their participation, showcasing differing levels of independence.

Recorded in the observers' research diaries (RT3), the construction process was reported as a positive experience for both caregivers and participants. The effective re-planning process contributes to a streamlined understanding, minimizing delays. For students with severe ASD P(F), the use of strategy with verbal and visual assistance was successful from the first action. By the last session, they managed to function as suppliers with visual prompts, as presented in Fig. 4.

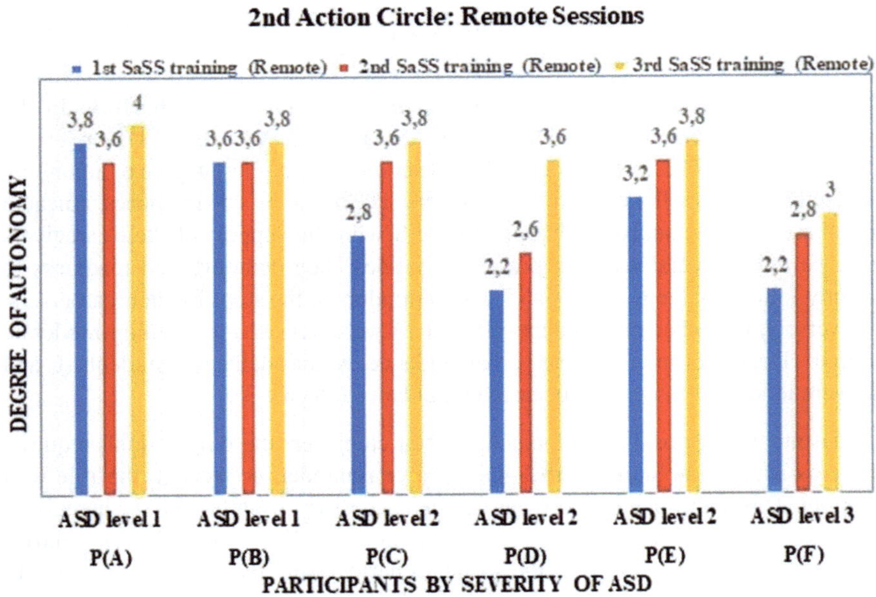

Fig. 4 Mean degree of autonomy in participation during the SaSS training as assessed in the second Action Circle by the independent observer

The remote SaSS training with caregiver support led to the autonomous participation of individuals with ASD in LEGO assembling activities as suppliers.

4.2 Findings Related to the Second RQ

With what techniques does SaSS enhance the participation of students with ASD in teamwork for assembling and programming educational robots?

Regarding the Findings of Face-to-Face, First Action Research Cycle In face-to-face sessions participants' challenges involved using LEGO Education WeDo 2.0 parts included in the kit, following the digital manual from LEGO Education WeDo 2.0 software. The team and coaches sat at a table. The coaches encouraged the students to share roles. They then introduced robotics materials to the students. They were reminded of the SaSS steps verbally and visually. The steps of the SaSS were visualized in the group so that students with typical development could learn what their role was and what the role of students with ASD was. Students were presented with an educational scenario reflection to trigger construction and programming.

According to the researchers' diaries (RT3), the students with the greatest need for support were the ones who mobilized the interest of students with typical development to help them participate. The coaches supported the students where needed but

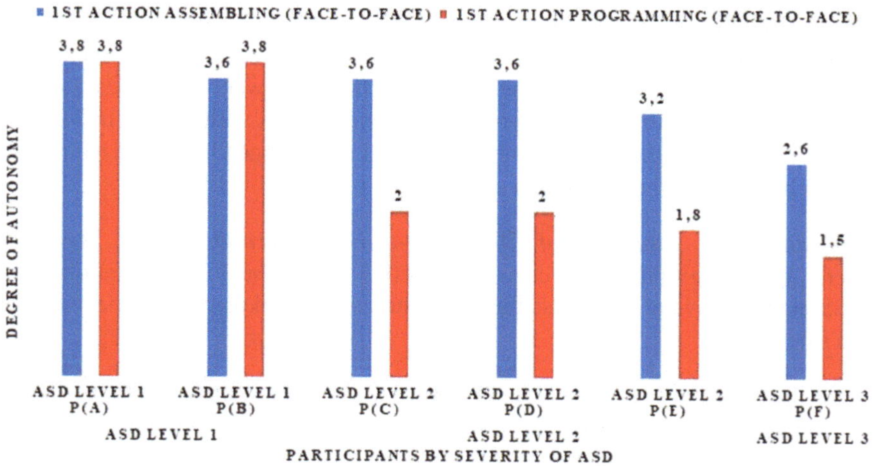

Fig. 5 Mean degree of autonomy in participation of the students with ASD during the educational robots assembling in an Inclusive Context with typical peers

gradually tried to reduce this to achieve independence. The structure of the model for step-by-step assembly as represented in the detailed manual proved to be an effective teaching material suitable for students with ASD and challenging simultaneously for typical students.

In assembling all the students except the one with ASD Level 3 participated occasionally supported by the typical peers or the researcher. This was impressive as the participants with the SaSS and the support of their peers managed to transfer in teamwork the SaSS and function as suppliers almost autonomously. Specifically, four students, P (A, B, C, D) managed to function almost autonomously as suppliers in teamwork with two additional peers. One student with ASD (F) participated with verbal support as his peers reminded the steps verbally and helped him to process the steps as a supplier in teamwork (Fig. 5).

In the programming phase, students with ASD Level 1 consistently maintained their level of autonomy using SaSS as suppliers of the coding blocks (Fig. 1). In more detail, as it is written in the diaries (RT3) the students (A) and (B) operated independently in their role as suppliers during the programming of LEGO Education WeDo 2.0 circular motion of the satellite and the forward-backward motion. They actively engaged in the design process, successfully locating and placing printed coding blocks in collaboration with their peers to insert them into the canvas and execute the code. However, most of the students (C, D, E, F) with ASD in Levels 2 and 3 during assembling functioned as suppliers without or with minimal support during the ER programming phase and demonstrated a need for more support with verbal and physical prompts. They engage in the programming process, requiring additional time to find and order the printed coding blocks for successful code execution.

Reflecting and Re-planning The SaSS underwent modifications to enhance the planning, aiming for a more streamlined implementation. The research team explored a linear SaSS design, where students with ASD Levels 2 and 3 retained the role of a "Supplier" of printed coding blocks on the canvas in the correct order and executed the program. This linear SaSS implementation aimed to empower neurotypical students to seek solutions with minimal involvement from students with ASD. The latter, in collaboration with other team members, dragged and dropped the organized printed coding blocks onto the canvas and executed the program.

Regarding the Findings of the Face-to-Face Second Action Research Cycle According to the researchers' diaries (RT3), the visual representation of the SaSS played a pivotal role in establishing a structured workplace with specific roles and rules.

As illustrated in Fig. 6 the ASD in Level 1 P(A) and (B) managed to participate effectively and autonomously in assembling and block programming of SaSS in block programming circular or forward and backward motion proved instrumental in enabling effective collaboration among students with ASD and their teammates, promoting inclusivity and equality. Students with Level 2 ASD (C, D, and E) were able to participate occasionally with peer support. The student with ASD Level 3 (F) was able to participate with constant verbal support from his peers. The interest of the student with ASD was maintained throughout the sessions. Correspondingly, the neurotypical students also considered the disabled student as a member of their group and reinforced him where necessary.

Summarizing the results from Stage II, the students demonstrated increased autonomy over sessions. After the reflection and react cycle, the SaSS was redesigned

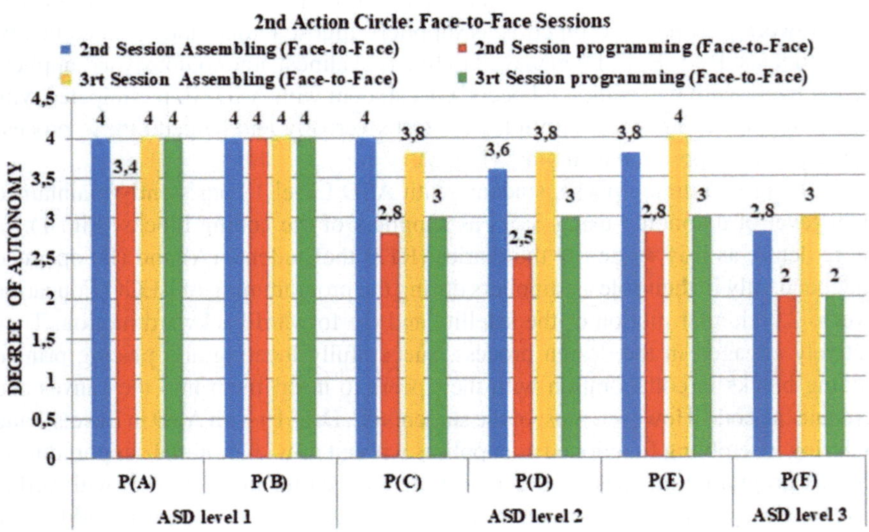

Fig. 6 Mean of SaSS usage assessment during the face-to-face inclusive educational robotics activities (ER), 2nd (face-to-face) action

for a more linear approach, empowering neurotypical students. The overall impact indicated increased autonomy during face-to-face activities.

Examining the overall impact, as presented in Fig. 8, the scores indicated a notable increase in autonomy during the latter. This observation reinforces the success of the research design, where students initially learned the strategy in remote sessions and implemented in face-to-face construction.

4.3 Findings Related to the Third RQ

What is the correlation between students' participation in remote SaSS training and their participation in assembly, and programming ER after training in SaSS? Can participation in one remote training predict participation in face-to-face ER?

The mean degree of autonomy in SaSS implementation is taught remotely with the support of the caregivers as illustrated in Fig. 7. Maintained face-to-face ER assembling with typical peers for all students.

As for block programming with WeDo 2.0 and Spike the students with ASD levels 2 and 3 (C, D, E, and F) managed to participate verbally and occasionally supported by their peers. Students with ASD implemented SaSS in block programming of the Satellite and Milo WeDo 2.0 and Spike robot to move forward and backward and arrange coding blocks in an accurate sequence, augmenting their enhancement in block-based coding activities printed coding blocks for left and right navigation of LEGO Education Spike (Fig. 8).

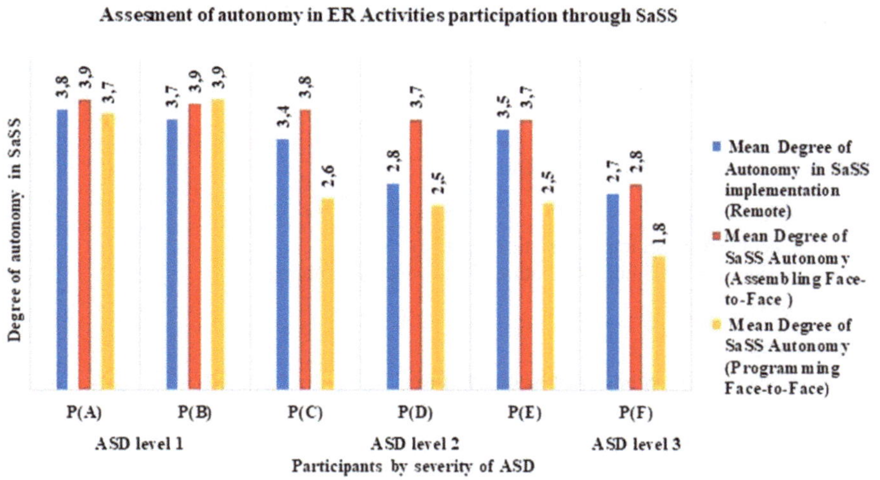

Fig. 7 Mean of SaSS implementation assessment

Fig. 8 Programming Spike Tools with printed coding blocks

As it is documented by the implementation of Pearson's Correlation Coefficient to measure the linear correlation, the variable (Degree of autonomy in SaSS implementation remotely) is in a linear correlation with the variable (Degree of autonomy in face-to-face in assembling), and with the variable (Degree of autonomy in face-to-face programming). The Pearson's Correlation Coefficient (r) detects a strong positive correlation (see Table 1).

Programming exhibited a gradual but consistent upward trend over the three in-person sessions. Data analysis suggests that students functioning at Level 1 achieved higher scores, approaching autonomous execution of the strategy. For students at Levels 2 and 3, stable scores indicated successful SaSS execution in programming with guidance from the group. Students with ASD (P E, P D) maintained autonomy in their roles throughout ER activities using the SaSS. Moderate ASD students (P A, P C, P F) demonstrated autonomy as suppliers in face-to-face construction activities but required occasional support in ER programming (Level 3). The student with severe ASD needed moderate support for construction and additional support with verbal instructions in programming.

Table 1 Pearson's correlation coefficient

Assembling
x = mean degree of SaSS implementation (remote) y = mean degree of SaSS implementation (face-to-face assembling)
$r(4) = 0.9831, p < 0.05$
Programming
x = mean degree of SaSS implementation (remote) y = mean degree of SaSS implementation (face-to-face programming)
$r(4) = 0.9793, p < 0.05$

4.4 Findings for the Fourth RQ

What are the views of typical peers and caregivers on student's participation in the program?

Regarding the fourth research question: Typical peers naturally took on the role of guides for students with disabilities, highlighting the significance of progress for everyone involved (RT4).

> For me, it is important not only for my process but the progress of all. It does not matter. I prefer to teach someone else and share knowledge.
>
> All the activities were very interesting! I enjoyed helping the team. I hope the program continues.

According to the caregivers' interviews and the recording of the emotions of students with ASD noted a positive outlook after face-to-face sessions (RT5).

> He was looking forward to the next session and expressed that he wanted to continue more.

Caregivers observed improved cooperation, enhanced fine motor skills, and increased socialization, underscoring the importance of parent training for effective participation. Caregiver feedback highlighted improved cooperation, fine motor skills, and socialization. Neurotypical developing students naturally took on the role of guides for student with ASD in face-to-face sessions.

At the end of each session, we gave each ASD student three cards that depicted three emotions: the first card had a smiley face showing a smiling face, the second a face showing a moderate emotion neither happy nor sad, and the third a sad face. We asked each student to choose the emotion that represented them at that particular moment for the activities we had just completed (RT6). In all three sessions, all students chose the emoticons that showed a happy emotion.

5 Discussion

The equal participation of all students, including those with ASD, in activities utilizing modern technologies in the context of typical and non-typical education, is a primary goal for fostering inclusive education and promoting social cohesion [17]. Current research on Inclusive Educational Robotics (IER) highlights the need to develop methodologies that enhance coding skills and the accessibility of ER interfaces thereby fostering the equal participation of students with ASD [10]. Following this direction, the present action research aimed to extend the Inclusive ER methodology based on SaSS training [9] to further develop its effectiveness. Specifically examined the remote extension of SaSS training investigating how could influence the participation of students with ASD in the face-to-face assembling and programming of educational robots in IERa with peers.

The results from the action research indicate that the remote SaSS training of students with ASD with their caregivers' support positively influenced their participation and autonomy in LEGO assembling using detailed manuals. Students showed gradual improvement reaching significant autonomy in LEGO assembling as "Supliers" in the final SaSS remote training sessions. The social scenarios used by caregivers in SaSS remote sessions proved crucial in facilitating students' participation, highlighting the importance of adapting the IER training methodologies to the educational needs of ASD students [5, 18].

In face-to-face IERa students with ASD, especially those at Levels 1 and 2, demonstrated significant autonomy when functioning as "Suppliers," supported by typical peers and coaches. The visual and verbal reminders helped maintain the focus of students with ASD and improved their autonomous participation in IER. Even the students with a need for a higher level of support (Level 3) managed to participate in face-to-face IER with visual prompts. The SaSS's structured, linear coding approach allowed neurotypical peers to lead problem-solving enabling students with ASD to contribute meaningfully and enhance their autonomy with inclusive teamwork.

The correlation analysis between the participation of students with ASD in remote SaSS training—supported by their caregivers—and their subsequent participation in face-to-face assembling and block programming in IERa revealed a strong positive relationship. Pearson's Correlation Coefficient analysis indicated that the mean degree of autonomy during remote SaSS training was significantly positively correlated with both face-to-face assembling and programming autonomy. These findings suggest that the skills and strategies acquired through remote SaSS training can effectively foster the participation of students with ASD in face-to-face IERa. Additionally, the degree of autonomy in the participation of students with ASD during remote SaSS training can be a reliable predictor of their autonomy participation in face-to-face IERa in assembling and block programming.

Typical peers took on the role of leaders, showing enthusiasm and satisfaction from participating and collaborating. Overall, SaSS training had a positive impact enhancing the participation and social skills of students with ASD but also for neurotypical students. The SaSS enabled students with ASD to significantly develop cooperation with typically developing students. It introduced them to their first contact with programming and coding skills.

6 Limitations

The limitation of our research concerns the number of participants. Confirmation of the effectiveness of SaSS requires investigating its application to a larger number of participants in typical education and non-typical education. Interviews with parents at the end of the program could be more specific. Finally, there were some difficulties in the remote sessions related to the weak internet connection, and the incorrect placement of the camera from caregivers, which made it difficult to accurately record the data.

7 Conclusions

The study suggests that remote SaSS training of students with ASD, with caregiver assistance, positively influences their autonomous participation in IERa. Moreover, the structured linear SaSS approach supports the development of coding skills on students with ASD creating a cooperative and inclusive learning environment. Given the positive findings, SaSS can be introduced into primary and secondary education, as a strategy that creates a structured, safe learning environment in order to enhance the equal participation of students with disabilities in inclusive educational robotic groups, enhancing a multitude of skills of all students. This requires appropriate training of teachers so that they know how to apply this specific strategy and will be flexible in applying appropriate techniques for its implementation. This research highlights the potential of SaSS training and underscores the need of further studies to validate its impact across diverse educational settings and student profiles. Our research team aims to further investigate the impact of SaSS training on a broader range of students with diverse disabilities.

References

1. Nanou, A., Tsiomi, E., Oikonomou, A., Karampatzakis, D.: The SAS strategy training for children with ASD in inclusive educational robotics activities. Educ. Innov. Div. **2**(3), 34–52 (2021). https://doi.org/10.17770/eid2021.2.6723
2. Caballero-Gonzalez, Y.-A., Munoz-Repiso, A.-G., Garcia-Holgado, A.: Learning computational thinking and social skills development in young children through problem solving with educational robotics. In: Proceedings of the Seventh International Conference on Technological Ecosystems for Enhancing Multiculturality, pp. 19–23. ACM, Spain (2019). https://doi.org/10.1145/3362789.3362874
3. Papazoglou, T., Karagiannidis, C., Mavropoulou.: Educational robotics can foster social inclusion and social status of children with autism. In: In: 2021 International Conference on Advanced Learning Technologies (ICALT), pp. 317–319. IEEE, Tartu (2021). https://doi.org/10.1109/ICALT52272.2021.00102
4. Fachantidis, N., Syriopoulou-Delli, C.K., Vezyrtzis, I., Zygopoulou, M.: Beneficial effects of robot-mediated class activities on a child with ASD and his typical classmates. Int. J. Dev. Disab. **66**(3), 245–253 (2019). https://doi.org/10.1080/20473869.2019.1565725
5. Koller, D., Stoddart, K.: Approaches that address social inclusion for children with disabilities: a critical review. Child Youth Care Forum. **50**(4), 679–699 (2021). https://doi.org/10.1007/s10566-020-09589-8
6. Benitti, F.B.V.: Exploring the educational potential of robotics in schools: a systematic review. Comput. Educ. **58**(3), 978–988 (2012). https://doi.org/10.1016/j.compedu.2011.10.006
7. Buchnat, M., Wojciechowska, A.: Online education of students with mild intellectual disability and autism spectrum disorder during the COVID-19 pandemic. Interdyscyp. Konteksty Pedag. Specjalnej. **29**, 149–171 (2020). https://doi.org/10.14746/ikps.2020.29.07
8. Tsiomi, E., Nanou, A.: Cooperative strategies for children with autism spectrum disorders in inclusive robotics activities. Soc. Integr. Educ. **4**, 148 (2020). https://doi.org/10.17770/sie2020vol4.5147
9. McCay, L.O., Keyes, D.W.: Developing social competence in the inclusive primary classroom. Child. Educ. **78**(2), 70–78 (2001). https://doi.org/10.1080/00094056.2002.10522707

10. Lewin, K.: Resolving Social Conflicts; Selected Papers on Group Dynamics. Harper, New York (1948)
11. Eisner, E., Eisner, E.W.: The Enlightened Eye: Qualitative Inquiry and the Enhancement of Educational Practice. Macmillan, New York (1991)
12. Kemmis, S., McTaggart, R., Nixon, R.: The Action Research Planner: Doing Critical Participatory Action Research. Springer, Singapore (2014)
13. Zhong, B., Li, T.: Can pair learning improve students' troubleshooting performance in robotics education? J. Educ. Comput. Res. **58**(1), 220–248 (2020). https://doi.org/10.1177/0735633119829191
14. Nanou, A., Karampatzakis, D.: The participation of students with autism in educational robotics: a scoping review. Soc. Sci. **12**, 675 (2023). https://doi.org/10.3390/socsci12120675
15. Hewitt, G., Sims, S., Harris, R.: Evidence of communication, influence and behavioural norms in interprofessional teams: a realist synthesis. J. Interprof. Care. **29**(2), 100–105 (2015). https://doi.org/10.3109/13561820.2014.941458
16. Nanou, A., Karampatzakis, D.: Collaborative educational robotics for the inclusion of children with disabilities. Educ. Innov. Div. **4**, 3043 (2022). https://doi.org/10.17770/eid2022.1.6899
17. Askari, S., Anaby, D., Bergthorson, M., Majnemer, A., Elsabbagh, M., Zwaigenbaum, L.: Participation of children and youth with autism spectrum disorder: a scoping review. Rev. J. Autism Dev. Disord. **2**, 103–114 (2015). https://doi.org/10.1007/s40489-014-0040-7
18. Le Goff, D.B.: Use of LEGO as a therapeutic medium for improving social competence. J. Autism Dev. Disord. **34**(5), 557–571 (2004). https://doi.org/10.1007/s10803-004-2550-0

Augmented Reality Educational Games for Disaster Management and Preparedness. The Case of Virtual School

Avgoustos Tsinakos, Ioannis Kazanidis, and George Terzopoulos

Abstract Disaster preparedness is a crucial aspect of ensuring the safety and well-being of communities, especially among vulnerable populations such as primary school students. To address this need, this research paper presents the development and evaluation of Virtual School, an Augmented Reality (AR)-based educational game designed to teach primary school students about natural disaster preparedness. In the current study we examined whether primary and high school students could learn disaster preparedness and response actions through the proposed game. The game allows students to collaborate and navigate through simulated disaster scenarios within a virtual school environment, acquiring essential knowledge and skills to respond effectively in real-world emergencies. An exploratory study measures perceptions of learning self-efficacy, enjoyment, and game usability. The results showed that players significantly improved their knowledge and performance while they have positive attitude over the proposed game.

Keywords Disaster preparedness · Serious games · Augmented reality · Game-based learning · Collaborative learning

1 Introduction

Natural disasters can have devastating consequences, particularly for vulnerable populations such as children in primary schools and disaster education needs to be developed by schools, especially those located in disaster-prone locations [1]. Disaster preparedness involves a series of proactive steps taken by governments,

A. Tsinakos (✉) · I. Kazanidis · G. Terzopoulos
Department of Informatics, Democritus University of Thrace, Kavala, Greece
e-mail: tsinakos@cs.duth.gr

I. Kazanidis
e-mail: kazanidis@cs.duth.gr

G. Terzopoulos
e-mail: gterzopo@cs.ihu.gr

I. Kazanidis and A. Tsinakos (eds.), *Research on E-Learning and ICT in Education*,
https://doi.org/10.1007/978-3-031-99961-1_14

organizations, communities, and individuals to improve their ability to respond to and manage the immediate consequences of a disaster. Usually, students receive information on how best to prepare for disasters through typical presentations, lectures, and slideshows. This method lacks interactivity and fails to engage students. Recently, researchers developed web-based games where students had to respond to simulated disaster scenarios. One example is a previous study by Blasko-Drabik et al. [2], where an online game was developed in order to assess students' confidence in their ability to take appropriate actions during natural disaster events. Although these game-based approaches aim to increase student engagement and acquisition of disaster preparedness knowledge and skills, due to their limited interactivity and immersion, they do not achieve sufficient levels of engagement and learning retention.

To address this, a promising solution seems to be the utilization of extended reality game-based applications. Immersive technologies combine the physical with the virtual world, allowing students to participate more active in simulated disaster scenarios [3]. According to Hsu et al. [4], these immersive applications can potentially enhance disaster preparedness by providing engaging and immersive learning experiences to students.

Games for disaster preparedness based on immersive technologies and mainly on Virtual Reality (VR) technology have been explored in the literature. Applications with gamification elements are shown to create new opportunities for community engagement in order to raise disaster awareness and build a "gamified culture" that supports smart disaster emergency planning [5]. Collaborative game-based learning platforms have also been developed in order to train students in post-disaster management and strategies, such as Rapid Damage Assessments and Incident Command System [6]. Moreover, VR technology is considered an important asset for training in disaster preparedness, offering trainees advanced tactical training in chaotic, simulated environments. In VR, trainees can practice continuously to master the required skills and techniques, which is not possible with traditional training methods [7].

Many of the aforementioned VR solutions require specific hardware infrastructures or focus primarily on post-disaster management and not on preparedness. Most of these systems are not specifically designed or can be adapted for use in primary or secondary education, limiting their effectiveness and accessibility for this specific target audience. Furthermore, VR technology has significant effects when used by primary school students, causing motion sickness, and raising concerns over safety and supervision, along with the need for specialized equipment. These can limit the accessibility and effectiveness of VR-based disaster preparedness training solutions for younger learners.

To address these challenges, a mixed approach, combining virtual and physical elements may be more suitable for engaging primary school students in disaster education. A promising approach is the use of an AR-based educational game with tangible interface that teaches primary school students about natural disaster preparedness. Compared to traditional training methods, this approach can engage students in a more effective and engaging way [8].

By integrating virtual and real-world components, the proposed approach can create an immersive learning environment that can focus the attention of young learners and provide a deeper understanding on disaster preparedness issues. Using Augmented Reality (AR) and gamification, students can actively participate in simulated disaster scenarios, solve real-world problems collaboratively, and apply incident management strategies in a safe and controlled environment. Benefits of augmented reality-based training have been highlighted in recent literature. Studies have demonstrated the effectiveness of AR-based training in improving emergency response skills and raising disaster awareness among communities. Extending this approach to include primary and high school context aims to build a strong foundation of preparedness among young learners, providing them knowledge and skills to respond effectively in case of a natural disaster event. However, the integration of such technologies into K12 education curriculums may also presents challenges. Challenges include ensured accessibility for all students, potential technical issues, and determining the most effective ways to incorporate AR experiences into the existing lesson plans. Thus, further research is needed to explore how AR experiences can be integrated effectively in primary education curriculums in order to enhance students' knowledge and preparedness for natural disasters.

This paper introduces the Virtual School, an AR-based educational game that allows primary school students to collaborate and find ways to escape from simulated disaster scenarios within their school environment. The game uses a real scale model of the school and AR-based augmentations to create an immersive learning experience. By immersing students in a combination of AR and tangible interface setting, this game-based approach aims to enhance their understanding of disaster preparedness, improve their response capabilities, and increase their resilience.

This paper will explore the design, development, and evaluation of the Virtual School game, highlighting its potential to serve as an effective tool for teaching natural disaster preparedness to primary school students. It will assess students' engagement, learning outcomes, and overall effectiveness in improving disaster preparedness.

2 Related Work

The use of extended reality technologies for disaster preparedness training has been explored in several studies. These studies have demonstrated the potential of immersive technologies to provide advanced tactical training in simulated chaotic environments, allowing trainees to continuously practice and master the required skills. For instance, a study by Freeman et al. [9] described the use of a VR patient simulation system for teaching emergency response skills.

However, the application of these technologies for disaster preparedness education targeting primary school students remains an underexplored area. One study by Herowati [1] examined the use of augmented reality-based media to improve disaster preparedness among junior high school students, highlighting the potential of these technologies to enhance students' understanding and application of disaster mitigation strategies.

Additionally, Abdulhalim et al. [10] found that immersive virtual reality serious games can be effective tools for enhancing the preparedness and response of different community groups, including the deaf and hard-of-hearing population, in the event of disasters such as earthquakes. Similarly other studies have explored the use of various immersive technologies for teaching disaster preparedness (over tornados) to high school students [11].

Another interesting approach is the "House of disasters" which is an escape from disaster game comprised by a scale model of a house connected with the Makey Makey toolkit and a computer program to create an interactive simulation of different disaster scenarios [12]. Students play collaboratively with the tangible interface of the house, interacting with the physical scale model and receiving appropriate sound notations to enhance the immersive experience. They work together to navigate the simulated disaster scenarios, identify potential hazards, and practice effective evacuation procedures within the tangible environment. Another educational game for primary school children is the "Emergency suitcase," which is a tangible interface in the form of a suitcase. In this game, students must find and select the appropriate items and tools to pack into the suitcase [13]. The game uses Makey Makey and Scratch programming to provide sound feedback to students' actions on the proposed scenarios. Furthermore, Kazanidis et al. [14] created an educational escape room game with various puzzles and scenarios. Students had to use multiple artifacts and provide correct answers to escape the room. This was found to be an engaging and effective learning tool for high school students.

While these studies have demonstrated the potential of combining tangible interfaces with digital notations for disaster preparedness education, there remains a lack of research focused specifically on the development and evaluation of AR-based educational games with tangible interfaces for teaching natural disaster preparedness to primary school students. Building on these previous efforts, this paper aims to investigate the potential of an AR-based educational game, the Virtual School, to enhance natural disaster preparedness among primary school students.

3 Virtual School Design and Development

The Virtual School was designed to address the abovementioned gap by creating an immersive AR-based educational game that engages primary school students in disaster preparedness training. The game combines a tangible scale model of a school (Fig. 1) with AR components to create an educational environment for disaster preparedness training.

The school scale model provides a tangible reference point for students to interact with, representing the school building and its surroundings. The scale model is designed to be interactive, allowing students to physically manipulate and explore the environment. For example, Playmobil figures representing students, furniture, and other obstacles are used to simulate the school layout. The tangible interface is combined with disaster scenario cards, each accompanied by three answer cards

Fig. 1 Virtual School scale model initial prototype

indicating the corresponding proposed actions for students to consider. These cards are strategically positioned throughout the scale model to guide students' responses.

Augmented reality overlays digital information, such as virtual objects, animations, and interactive elements, onto the physical scale model, creating an immersive and engaging learning experience. Students can interact with the 3D model of the school and the AR content, exploring the school, identifying potential hazards, and practicing evacuation routes and emergency response procedures. They use their mobile devices to scan the answer cards, triggering the display of interactive elements and digital information overlays. These overlays provide additional guidance, explanations, and feedback to the students as they navigate the disaster scenarios. The augmentations were built using the free-to-use educational augmented reality platform ARTutor [15], and they feature video animations tailored to the age of primary school students. No programming skills were required to create the augmentations.

The applied scenarios simulate earthquakes, floods, fires, and storms, and students are asked to respond appropriately based on their location within the school, such as in a first-floor/ground-floor classroom, the yard, or other areas typical of a school building. The game incorporates a collaborative aspect, where students work in teams to navigate the scale model, identify hazards scenarios, and decide on the best course of action. Teachers are provided with analytical guide for the school scale model construction, and with lesson plans and curriculum materials that can be integrated into their existing teaching practices.

3.1 Virtual School Activity and Gameplay

The Virtual School activity begins with the construction of the school model based on analytical instructions and ready to print scenario cards.

Teachers first introduce the topic of natural disasters and their potential impact on schools to the students. They provide an overview of the Virtual School game, its objectives, and the importance of disaster preparedness. Students are then divided into small teams and assigned to specific disaster scenarios.

Each team is tasked with exploring the school model, identifying the relevant hazards, and selecting the appropriate response actions from the provided cards. In the Virtual School game, students will be able to select from various disaster scenarios relevant to a school setting, such as a fire in the classroom or an earthquake. Upon selecting a scenario, students receive information about the simulated emergency situation. For example, in the case of a classroom fire, students will be alerted to the location of the fire and any other relevant details about the situation.

Students will then be prompted to propose three alternative escape routes from the dangerous situation. For example, a card could mention that students should go to the ground floor using the elevator in case of fire. This card should be positioned in front of the elevator. Students should choose the correct answer and place their Playmobil figure in front of the appropriate card. After proposing their escape routes, students will receive feedback on their choices. The ARTutor app will analyze students' proposed routes and provide feedback on their safety and effectiveness. For example, if students propose an escape route that would lead them closer to the fire, the ARTutor app could alert them to the danger and suggest safer alternatives. The teams then discuss and present their findings to the class, sharing their reasoning and the lessons learned.

The teacher observes and guides the students throughout the activity, encouraging collaboration and discussion among the teams. After the initial exploration, the teams rotate to different disaster scenarios, building upon their previous experience.

4 Methodology

To evaluate the effectiveness of the Virtual School game in enhancing primary school students' natural disaster preparedness, an exploratory study was conducted that includes a pilot phase for testing and refining the game, followed by a larger-scale main phase to assess its broader impact, and qualitative data from a focus group for deeper insights.

The first phase involved a pilot study with a small group of 5 primary school students to gather preliminary feedback on the game design, usability, and learning outcomes. During this phase, researchers observed the students' interactions with the game, collected their comments and suggestions, and conducted pre- and post-assessment surveys to measure changes in their knowledge and attitudes related to disaster preparedness.

The larger-scale main phase was conducted in 10 primary and high schools to further validate the game's effectiveness. In this phase, researchers recruited a larger sample of students and conducted comprehensive assessments to measure the impact of the Virtual School game on students' disaster preparedness knowledge and self-efficacy. Details of the sample, the instruments and data collection, and the procedure regarding the second phase are presented in the next sections.

4.1 Sample

The second phase of the study prioritized the participation of isolated and remote schools in Greece and Cyprus, including those located in geographically challenging areas (distant Greek small islands) or with limited access to educational resources. This ensured a diverse representation of primary school students, especially those who may face unique challenges in preparing for and responding to natural disasters. The target group were students 9–11 years old (of fifth and sixth grade primary school students) and of 11–14 years old (high school students) from 10 schools all over Greece and Cyprus. A total of 94 students participated in the Virtual School game activity. All students completed the pre-test questionnaire, while 66 of them voluntarily answered the post-test questionnaire. Finally, schools selected a designated teacher to run the pilot implementation with the students.

4.2 Instruments and Data Collection

This research utilized a mixed-methods approach, combining quantitative and qualitative data collection and analysis. The assessments included pre-and post-game knowledge tests to measure students' understanding of natural disaster risks, evacuation procedures, and appropriate response strategies. Additionally, a survey was administered to assess students' perceived self-efficacy in dealing with disaster situations after playing the game, as well as their attitudes toward the game and the applied technologies.

The pre-and post-game knowledge tests included 12 open-ended questions to evaluate the students' knowledge, critical thinking skills, and problem-solving abilities related to natural disaster preparedness and response. The questions covered a range of topics, including identifying potential hazards, selecting appropriate evacuation routes, and demonstrating effective response strategies in the event of various natural disasters such as earthquakes, floods, fires, and storms. Additionally, the assessment included questions on the emergency contact numbers, such as the three-digit phone numbers for the Police, Fire Department, National Center for Emergency Aid, and Coast Guard Emergency Response, in each country.

The perceived self-efficacy and students' attitudes survey, asked students to rate their confidence and ability to respond to different disaster scenarios using a 7-point Likert scale. Additionally, the attitudes questionnaire assessed their opinion over the game and the usefulness of AR technology in disaster preparedness lessons based on Technology Acceptance Model (TAM) [16].

Finally, the focus group discussions and teacher interviews explored the participants' overall experience with the Virtual School game, its perceived impact on learning outcomes, and their feedback on the game's design, usability, and potential for integration into the school curriculum.

4.3 Procedure

We followed a structured implementation process. First, the research team engaged in workshops with participating schools to introduce the Virtual School game and provide training on its use. Teachers were given detailed guides and support materials to facilitate the implementation of the game within their classrooms.

Initially, they conducted a pre-assessment to measure students' baseline knowledge and self-efficacy. Students then worked in teams to construct the Virtual School scale model and printed the scenarios cards through a multiple session approach. They placed Playmobil figures into the school model and decorated accordingly. Then, students played the game in teams of four or five students (Fig. 2). Throughout the gameplay sessions, students engaged in collaborative problem-solving, navigating through the disaster scenarios and making decisions on appropriate response actions. Teachers closely observed the students' interactions, collected valuable feedback, and provided guidance to the students to ensure a smooth and productive gameplay experience.

After the gameplay sessions, students completed post-assessment surveys to measure changes in their disaster preparedness knowledge and self-efficacy, as well as their attitudes toward the game and applied technologies. Additionally, an online focus group was conducted with teachers, to gather their feedback and perspectives on the experience (Fig. 3).

5 Results

Preliminary findings from the pilot study that took place with five students during the first phase of the study, indicated that students who played the Virtual School game showed significant improvements in their disaster preparedness knowledge, and they

Fig. 2 Students playing the virtual school game

Fig. 3 Teachers online focus group

also reported increased confidence in their ability to respond effectively during emergency situations. Additionally, during the first phase, minor improvements needed for the augmentations and the school model construction details were implemented.

The larger-scale study (second phase) involving 94 students from 10 primary and higher education level schools across Greece and Cyprus showed similar positive results.

In particular, students' initial knowledge scores were on average 42.5% in the pre-test. After playing the Virtual School game, students' knowledge scores increased by an average of 46.9% points, reaching 89.4% (SD 0.66) in the post-test. At Fig. 4, the percentage of pre and post-test results on students' knowledge per disaster categories are shown. The learning gain rates from 35.4 for flood disasters to 53.3 for knowing emergency phone numbers.

Self-efficacy survey showed that students' perceived ability to respond effectively to disaster scenarios increased significantly, with an average score of 6.1 out of 7 after playing the game. This is translated to a rate of 89.7% perceived self-efficacy for dealing with natural disaster scenarios (Fig. 5).

The quantitative data about the TAM model revealed that students found both easy (5.5) and useful (5.8) the proposed game and technologies for the instruction of disaster preparedness. Particularly, students reported an average of 5.8 out of 7 for perceived usefulness and 5.5 out of 7 for ease of use. They also stated that they would like to see this type of AR-based educational games based on collaboration integrated into schools' curriculum to enhance learning in other domains as well (6.1).These findings align with previous research indicating that interactive learning methods for emergency preparedness are more effective than traditional approaches relying on written materials and lectures [14, 17, 18].

The qualitative data gathered through focus groups and teacher interviews provided additional insights into the strengths and potential improvements for the

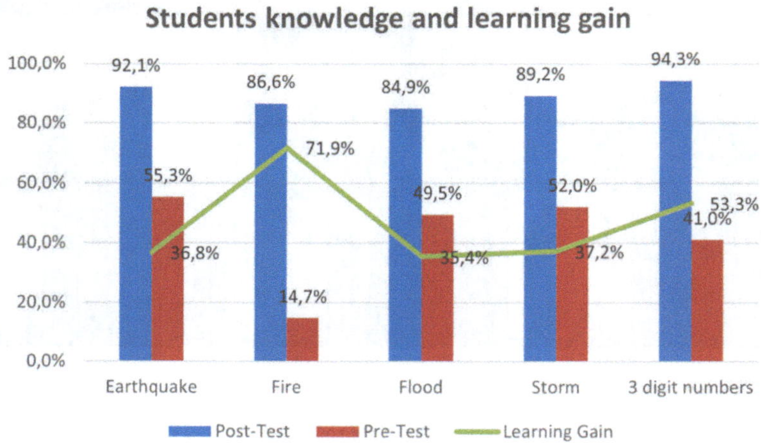

Fig. 4 Students' knowledge increase per disaster scenario after playing the virtual school game

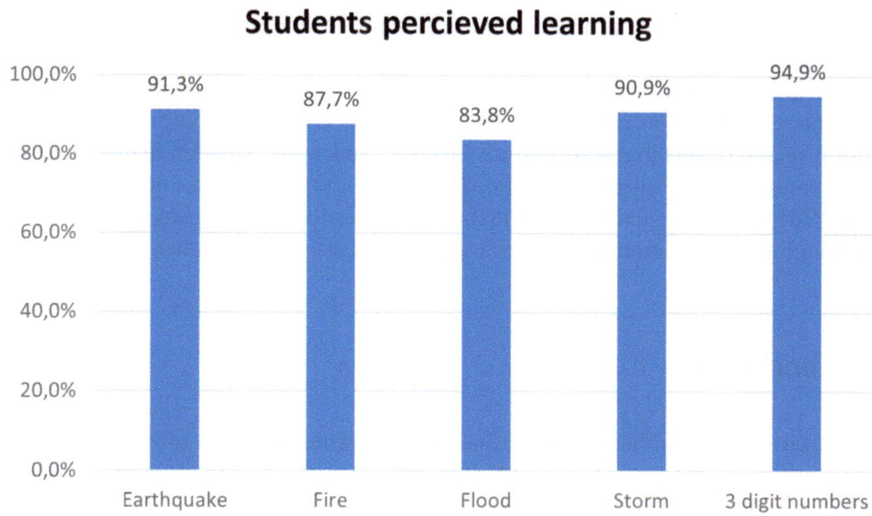

Fig. 5 Percentage of perceived learning per disaster scenario

Virtual School game application. Students reported that they feel more confident to respond to natural disasters and that the extended reality experience and collaborative problem-solving aspects of the game resulted in making the learning process more enjoyable and memorable. They also felt excited to the potential of the approach to be adopted more widely in their education.

Teachers also reported on the positive impact on student engagement and learning and highlighted the value of the approach in promoting critical thinking, collaboration, and twenty-first-century skills. In order to seamless integrate the Virtual School game into their curricular activities, teachers expressed the need for teacher training and additional supporting material.

6 Discussion

This paper proposes Virtual School, an AR-based educational game that combines physical and digital elements, an effective tool for teaching primary and secondary school students regarding natural disaster preparedness [2, 5]. Our findings suggest that there are significant improvements in students' disaster preparedness knowledge and self-efficacy after playing the game [2, 19]. Moreover, there is potential for wider adoption of the game since high levels of perceived usefulness and ease of use of the game and the associated technologies were reported. Student engagement and feedback was positive, and students reported increased confidence and enjoyment during the learning process. These findings are in consistency with previous studies [13, 14]. The success of the Virtual School game shows that immersive and collaborative learning approaches using emerging technologies to engage students in learning experiences have significant value.

Teachers' observations also highlight the value in promoting twenty-first-century skills like collaboration, critical thinking, and problem-solving. Thus, results confirm prior research results that demonstrate the effectiveness of extended reality environments for science and disaster education [19, 20]. The collaborative, interactive nature of the Virtual School game aligns with best practices for engaging students in active, meaningful learning around complex topics like disaster preparedness.

7 Conclusion

The study findings indicate that Virtual School game can be an effective learning tool for primary and secondary school students who want to learn more about natural disasters and their preparedness. This study demonstrates that game-based applications can enhance the disaster preparedness knowledge and the self-efficacy of the participating students, as well promote engagement, enjoyment, and the development of critical twenty-first-century skills.

Nevertheless, there are several limitations to the study. Although adequate for an initial pilot, the sample size was small and context-specific. These findings would require further validation with a larger sample and broader student population. In addition, the study was limited to the short-term effects of a Virtual School game and did not assess the retention of knowledge and acquired skills through the experience over a longer period of time. Future research needs to assess if the learning gains are sustained over the long term. In addition, despite the positive qualitative feedback we received more in-depth analysis of the student and teacher experiences could provide richer insights to inform future iterations of the game design and implementation strategy. Despite these limitations, the overall findings of this study suggest a promising role for AR-based gaming approaches for delivering disaster preparedness education.

The Virtual School game could be a suitable resource to include as a component of disaster preparedness curricula, as it promises to be an engaging approach for teaching important knowledge and skills. Considerations for broader implementation include ensuring adequate teacher training and support, as well as integrating the game seamlessly within existing lesson plans and activities. Overall, this study contributes to the growing body of evidence supporting the use of immersive, game-based learning to address complex educational challenges and prepare the next generation for the realities they may face in the future.

Acknowledgments We would like to express our gratitude to the undergraduates students Christoforakis C., Deligiorgi E., Konsoula E., Masali E. and Tektonidou C. who developed and tested the prototype school scale model.

Disclosure of Interests The authors have no competing interests to declare that are relevant to the content of this article.

References

1. Herowati, D.K.: Augmented reality-based media to improve disaster preparedness for junior high school students. In: Presented at the 5th International Conference on Current Issues in Education (ICCIE 2021). ICCIE, Yogyakarta (2022). https://doi.org/10.2991/assehr.k.220129.030
2. Blasko-Drabik, H., Blasko, D.G., Lum, H.C., Erdem, B., Ohashi, M.: Investigating the impact of self-efficacy in learning disaster strategies in an on-line serious game. Proc. Hum. Factors Ergon. Soc. Annu. Meet. **57**, 1455–1459 (2013). https://doi.org/10.1177/1541931213571325
3. Gampell, A., Gaillard, J.C., Parsons, M., Le Dé, L.: 'Serious' disaster video games: an innovative approach to teaching and learning about disasters and disaster risk reduction. J. Geogr. **119**, 159–170 (2020). https://doi.org/10.1080/00221341.2020.1795225
4. Hsu, E.B., Li, Y., Bayram, J.D., Levinson, D., Yang, S., Monahan, C.: State of virtual reality based disaster preparedness and response training. PLoS Curr. (2013). https://doi.org/10.1371/currents.dis.1ea2b2e71237d5337fa53982a38b2aff
5. Kankanamge, N., Yigitcanlar, T., Goonetilleke, A., Kamruzzaman, M.: How can gamification be incorporated into disaster emergency planning? A systematic review of the literature. Int. J. Disaster Resil. Built Environ. **11**, 481–506 (2020). https://doi.org/10.1108/IJDRBE-08-2019-0054
6. Babu, S.K., McLain, M.L., Bijlani, K., Jayakrishnan, R., Bhavani, R.R.: Collaborative game based learning of post-disaster management: serious game on incident management frameworks for post disaster management. In: 2016 IEEE Eighth International Conference on Technology for Education (T4E), pp. 80–87. IEEE, Mumbai (2016). https://doi.org/10.1109/T4E.2016.024
7. Eva, P., Ladislav, H.: Virtual reality as needful factor of intervention in natural disasters. In: 2017 International Conference on Engineering, Technology and Innovation (ICE/ITMC), pp. 1–5. IEEE, Funchal (2017). https://doi.org/10.1109/ICE.2017.8279861
8. Arinta, R.R., Suyoto, S., Emanuel, A.W.R.: Effectiveness of gamification for flood emergency planning in the disaster risk reduction area. Int. J. Eng. Pedag. **10**, 108 (2020). https://doi.org/10.3991/ijep.v10i4.13145
9. Freeman, K.M., Thompson, S.F., Allely, E.B., Sobel, A.L., Stansfield, S.A., Pugh, W.M.: A virtual reality patient simulation system for teaching emergency response skills to U.S. Navy Medical Providers. Prehosp. Disaster Med. **16**, 3–8 (2001). https://doi.org/10.1017/S1049023X00025462
10. Abdulhalim, I., Mutch, C., González, V.A., Amor, R.: Improving post-earthquake evacuation preparedness for deaf and hard of hearing children: a conceptual framework. Int. J. Disaster Risk Reduct. **62**, 102360 (2021). https://doi.org/10.1016/j.ijdrr.2021.102360
11. Hiroyuki, M., Masami, S.: Comparative experiments on simulated tornado experience via virtual reality and augmented reality. Inf. Syst. Educ. **19**, 21–31 (2020)
12. Kazanidis, I., Panagiotopoulou, K., Palaigeorgiou, G., Tsinakos, A.: The house of disasters: learning to deal with disasters through tangible interfaces. In: Proceedings of the 6th Panhellenic Conference "Integration and Use of ICT in the Educational Process, pp. 54–65 (2019)
13. Panagiotopoulou, K., Kazanidis, I., Tsinakos, A.: The construction of a physical interface for educating children on disaster management, using Scratch and Makey-Makey: The emergency suitcase. In: Proceedings of the 5th Panhellenic Conference Integration and Use of ICT in Education, pp. 582–593 (2017)
14. Kazanidis, I., Gotzamanis, V., Fotaris, P., Tsinakos, A.: Educational escape room for disaster preparedness and response training. In: Proceedings of the 14th European Conference on Game Based Learning, ECGBL 2020, pp. 832–839 (2020). https://doi.org/10.34190/GBL.20.093
15. Terzopoulos, G., Kazanidis, I., Tsinakos, A.: Building a general purpose educational augmented reality application: the case of ARTutor. In: Auer, M.E., Tsiatsos, T. (eds.) New Realities, Mobile Systems and Applications, pp. 168–179. Springer, Cham (2022). https://doi.org/10.1007/978-3-030-96296-8_16
16. Davis, F.D.: Perceived usefulness, perceived ease of use, and user acceptance of information technology. MIS Q. **13**, 319 (1989). https://doi.org/10.2307/249008

17. Novak, J., Lozos, J.C., Spear, S.E.: Development of an interactive escape room intervention to educate college students about earthquake preparedness. Nat. Hazards Rev. **20**, 06018001 (2019). https://doi.org/10.1061/(ASCE)NH.1527-6996.0000322
18. Becker, J.S., Paton, D., Johnston, D.M., Ronan, K.R.: A model of household preparedness for earthquakes: how individuals make meaning of earthquake information and how this influences preparedness. Nat. Hazards. **64**, 107–137 (2012). https://doi.org/10.1007/s11069-012-0238-x
19. Smith, T.: Elementary science instruction: examining a virtual environment for evidence of learning, engagement, and 21st century competencies. Educ. Sci. **4**, 122–138 (2014). https://doi.org/10.3390/educsci4010122
20. Caroca, J., Bruno, M.A., Aldunate, R.G., Arancibia, C.U.: Epistemic video game for education in wildfire response: a pilot study. Int. J. Technol. Enhanc. Learn. **11**, 247 (2019). https://doi.org/10.1504/IJTEL.2019.100481

Unlocking the Potential of Immersive Technologies in K-12 Physical Education: A Systematic Literature Review

Evaggelia Andreadou and Ioannis Kazanidis

Abstract Immersive technologies such as augmented reality (AR) and virtual reality (VR) offer innovative opportunities to enhance physical education (PE) by promoting physical, cognitive, and social growth. This systematic literature review examines their integration into K-12 PE curricula, analyzing 33 peer-reviewed studies published between 2012 and 2022, following PRISMA guidelines. The findings highlight that immersive technologies significantly improve motor skills, cognitive abilities, and social interaction while increasing student engagement, motivation, and enjoyment in PE classes. Students benefit from enhanced learning outcomes, deeper participation, and innovative experiences, though challenges such as reduced physical activity in specific contexts and implementation barriers persist. Educators are encouraged to adopt these technologies through accessible tools and professional training to maximize their potential. Further research is needed to explore long-term effects, cost-effectiveness, and best practices for incorporating immersive technologies into diverse educational settings. This review underscores the transformative potential of AR and VR to revolutionize K-12 PE, providing actionable insights for designing effective, technology-driven curricula.

Keywords Augmented reality (AR) · Virtual reality (VR) · Immersive technologies · Physical education · Primary education · Exergames

1 Introduction

There are a number of barriers to children's and adolescents' physical activity, such as obesity, poor self-image due to ineffective parenting, or parents' overprotection, aiming at preventing their children from exposure to potential risks [1]. Moreover,

E. Andreadou (✉)
Department of Informatics, Democritus University of Thrace, Kavala, Greece
e-mail: andreadoue@gmail.com

E. Andreadou · I. Kazanidis
International Hellenic University, Thermi, Greece
e-mail: kazanidis@cs.duth.gr

I. Kazanidis and A. Tsinakos (eds.), *Research on E-Learning and ICT in Education*,
https://doi.org/10.1007/978-3-031-99961-1_15

their lifestyles have become more sedentary than ever, mainly due to increased screen time, which affects their physical and mental health [2] and causes or contributes to children's and adolescents' inactivity, which is among the main causes of obesity, diabetes, and cardiovascular disease [3].

However, it has been documented that virtual reality (VR) and augmented reality (AR) technologies can facilitate and improve the processes of practice and training [4]. Although using their smart devices on a regular basis does not seem optimal, the educational benefits seem to counterbalance any reservations [5], providing Physical Education (PE) teachers with the opportunity to use immersive technologies in various learning contexts [6].

Research on the integration of immersive technologies in education has revealed positive outcomes regarding students' attitudes, motivation, engagement, skills development or improvement, and performance, especially when compared with traditional methods [4, 7], mainly because students are enabled to have first-person experiences [8]. As a result, AR and VR technologies enrich the learning process, promote communication and collaboration among students, offer deep learning opportunities [7], improve all twenty-first-century skills, and provide engaging learning environments that facilitate learning and affect the cognitive, intrapersonal, and interpersonal domains.

Although educational policies around the world have included digital tools as a means of supporting teaching and learning [4], there is limited research on the potential and contribution of immersive technologies within the context of PE [2–4, 9–13] leaving a considerable gap in the literature. Moreover, it has been observed that the majority of the studies explore the effect of technology on student motivation, fitness, motor skills, physical activity, and cognitive motivation or understanding, without considering the learning outcomes of using immersive technologies in a comprehensive manner, examining physical, cognitive, social, affective, and attitudinal benefits [10], as the present study aims to do. This paucity is even more obvious when considering the rapid advances of technology, revealing the need for further exploration [3].

To address the identified gaps in the existing literature, this systematic review seeks to consolidate current evidence on the benefits of immersive technologies in K-12 PE. Specifically, this study aims to elucidate the use and advantages of Augmented Reality and Virtual Reality technologies in enhancing PE outcomes. Given the significant interest in the potential of immersive technologies across various educational domains, it is crucial to investigate their impact within the context of K-12 PE. This review addresses the following research questions:

- RQ1. What are the physical benefits of the use of immersive technologies in PE gained by K-12 students?
- RQ2. What are the cognitive benefits of the use of immersive technologies in PE gained by K-12 students?
- RQ3. What are the psychological benefits of the use of immersive technologies in PE gained by K-12 students?

- RQ4. What are the social benefits of the use of immersive technologies in PE gained by K-12 students?
- RQ5. What are the attitudinal benefits of the use of immersive technologies in PE gained by K-12 students?

The remainder of this article is organized as follows. Section 2 outlines the methodology used in this systematic review, including the inclusion and exclusion criteria, search strategy, and data extraction process. Section 3 presents the results of the review, categorized into physical, cognitive, psychological, social, and attitudinal benefits of immersive technologies in K-12 PE. Section 4 discusses these findings, highlighting the implications for practice and potential limitations of the current research. Finally, Section 5 offers conclusions and suggests directions for future research, emphasizing the need for longitudinal studies and broader implementation of AR and VR in diverse educational settings.

2 Materials and Methods

To comprehensively address the research questions, this systematic literature review was conducted following the Preferred Reporting Items for Systematic Reviews and Meta-Analyses (PRISMA) guidelines. The review included peer-reviewed journal articles published between 2012 and 2022, reflecting the rapid advancements in immersive technologies. Studies were selected based on predefined inclusion and exclusion criteria to ensure the relevance and quality of the data. This approach enabled a focused examination of the impacts of AR and VR technologies in K-12 Physical Education settings.

2.1 Inclusion and Exclusion Criteria

This review included studies that met specific criteria to ensure the analysis reflects recent advancements and maintains high research standards. Articles published between 2012 and 2022 were considered, focusing on the latest developments in immersive technologies. Only peer-reviewed journal articles were included to ensure the reliability and quality of the data. Additionally, the review was limited to studies written in English to ensure accessibility and comprehension.

The studies needed to focus on the use of Augmented Reality or Virtual Reality in K-12 Physical Education settings, involving primary and secondary education students. This focus ensured that the findings were directly relevant to the target educational contexts.

Conversely, several exclusion criteria were applied to maintain the integrity and focus of the review. Non-peer-reviewed publications, including theoretical papers, meta-analyses, dissertations, and conference proceedings, were excluded to concentrate on empirical studies with robust methodologies. Articles published before 2012 were also excluded to avoid outdated insights and ensure the review's relevance to current educational practices. Studies that did not focus on PE or involved subjects other than K-12 students were also excluded to maintain a clear and specific focus on the intended research questions.

2.2 *Identification of Studies*

The search for relevant literature was exclusively conducted using Google Scholar, chosen for its comprehensive repository and extensive indexing of peer-reviewed online journals. This platform's broad coverage ensures access to a wide range of academic articles. The search strategy employed combinations of the following keywords: "("augmented reality" OR AR) AND ("virtual reality" OR VR) AND "exergames" AND ("physical education" OR PE) AND (K-12 OR "primary education" OR "elementary school" OR education)".

2.3 *Study Selection and Data Extraction*

A total of 964 results were retrieved through the database search. Duplicate articles were removed, yielding 947 unique records for screening. Many of the records retrieved did not involve student participants and thus were outside the scope of this review, including articles not related to immersive technologies. After reviewing the titles and abstracts, we assessed 62 full-text articles for eligibility. Of these, 29 articles were excluded based on the inclusion and exclusion criteria: Finally, 33 articles were included in this review. The systematic review PRISMA flow diagram is presented in Fig. 1.

Data management was facilitated through the creation of a Microsoft Excel spreadsheet, organized with columns labeled: Name(s), Year, Research Method, AR/VR Technology, Setting, Learning Theory/Approach, Conditions, Benefits, Difficulties/ Challenges, Limitations, and Conclusions. Each study was thoroughly analyzed and detailed within these categories by the two authors. During this process, three of the selected articles were excluded, due to insufficient reporting of essential information or significant methodological flaws that precluded comprehensive data entry for all specified categories.

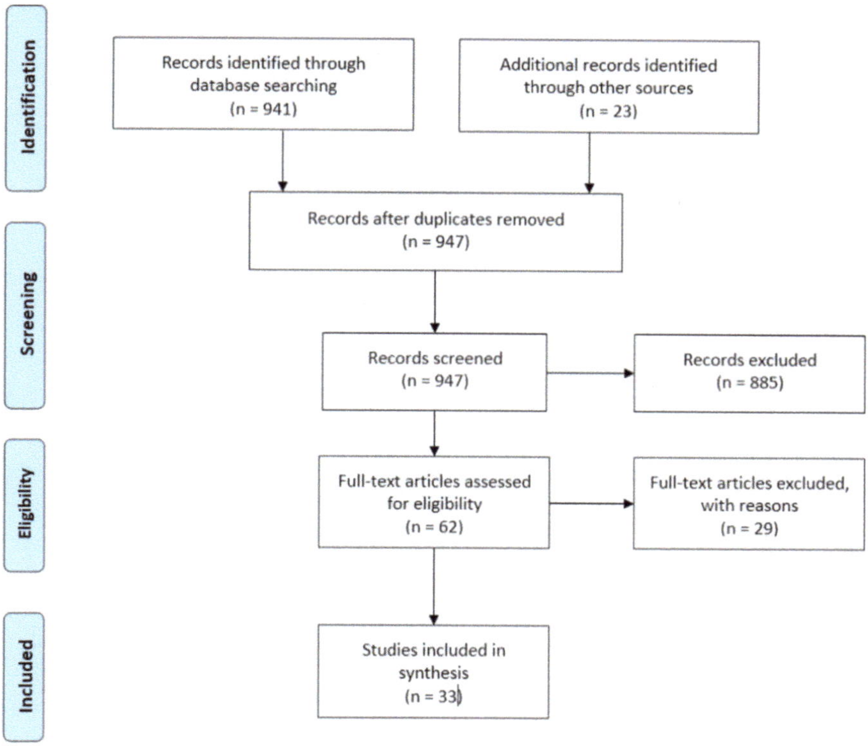

Fig. 1 PRISMA 2020 flow diagram

2.4 Data Analysis

Upon completing the quality assessment, the data from the selected studies were systematically coded and synthesized to align with the research questions. This involved categorizing the benefits of using immersive technologies in K-12 Physical Education into broader themes: physical, cognitive, psychological, social, and attitudinal. Each study was reviewed, and relevant data were extracted and entered into the pre-structured Excel spreadsheet. Descriptive statistics were used to summarize the characteristics of the included studies, such as sample size, duration, and the particular AR/VR technologies utilized.

Thematic analysis was employed to uncover and examine patterns within the data, allowing for a comprehensive understanding of the diverse effects of AR and VR on K-12 PE. Studies were grouped based on the type of benefits reported, and the frequency of each benefit was assessed to identify the most commonly observed outcomes. This classification of benefits is visually visual depicted in Fig. 2, which illustrates the distribution and frequency of physical, cognitive, psychological, social, and attitudinal benefits reported across the analyzed studies.

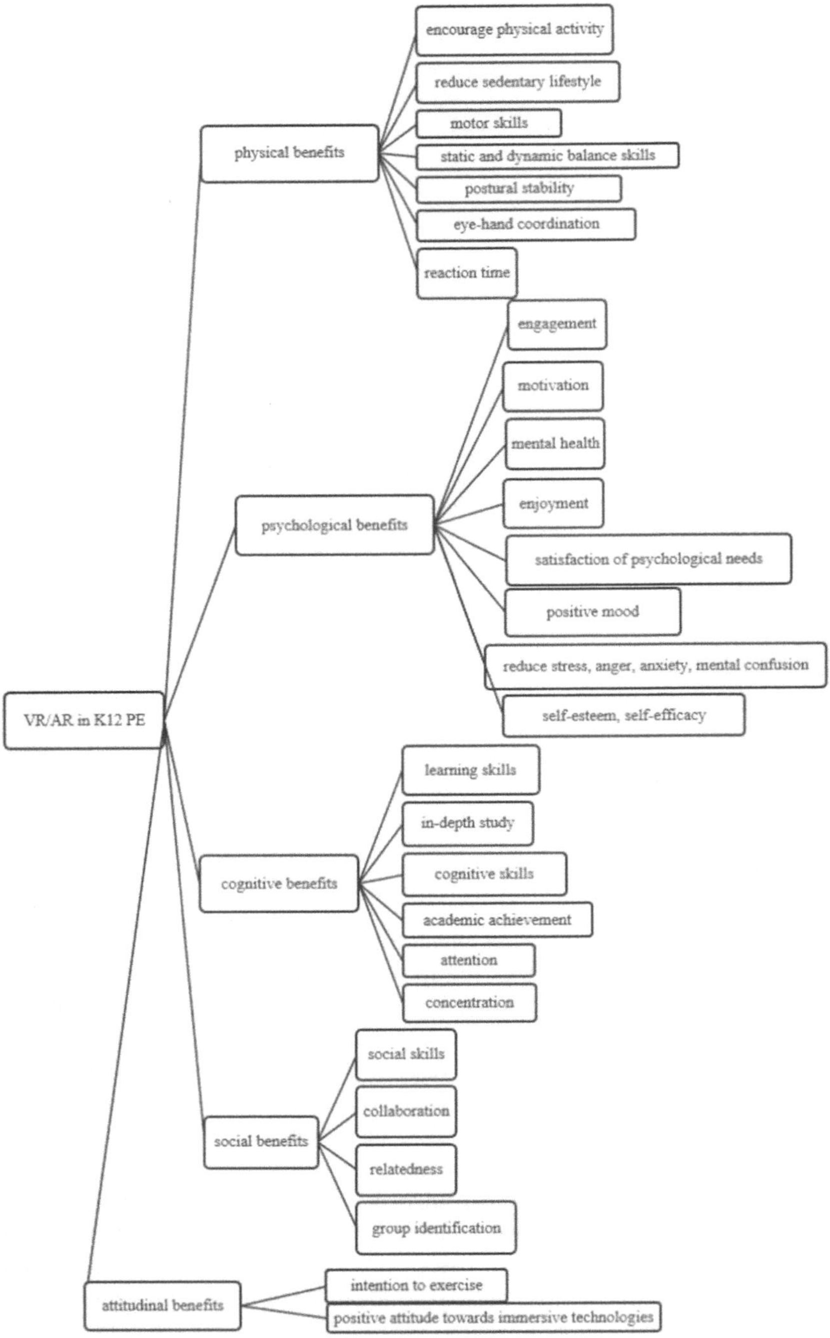

Fig. 2 Distribution of benefits from AR and VR in K-12 physical education

Furthermore, any noted difficulties, challenges, and limitations were also documented to provide a well-rounded perspective on the implementation and impact of these technologies. This thorough data analysis ensured that the review captured both the positive effects and the potential drawbacks of using AR and VR in K-12 PE, offering a nuanced insight into their significance in educational environments.

3 Results

3.1 Key Attributes of Included Studies

To provide a comprehensive understanding of the studies included in this review, Table 1 summarizes their key attributes. This table includes information such as the names of the author(s), the year of publication, the sample sizes and demographics, the instructional settings, and the duration of the experiments.

The range of publication years from 2012 to 2022 reflects the rapid advancements in immersive technologies and their application in K-12 PE. The sample sizes varied significantly, encompassing a broad spectrum of student populations from different educational levels and geographical regions. Instructional settings included classrooms, gyms, and specialized labs, highlighting the versatility of AR and VR technologies in various PE environments. The duration of the studies also varied, from short-term interventions lasting a few weeks to longitudinal studies spanning several months, providing insights into both immediate and long-term impacts of immersive technologies.

3.2 Methodological Procedures

As shown in Fig. 3, most studies in this review used experimental designs with quantitative methodologies, highlighting the focus on objective measurements in assessing the impacts of immersive technologies in K-12 PE. Quantitative methods allowed for evaluating changes in physical activity levels, cognitive performance, and other measurable outcomes.

A few studies utilized different approaches. Finco et al. [22] conducted the only qualitative study, offering insights into participants' experiences. Quintas-Hijós et al. [38] used a mixed methods approach, combining qualitative and quantitative data for a comprehensive analysis. Meckbach et al. [20] also employed mixed methods, including video observation alongside quantitative data. In terms of duration, Ruiz-Ariza et al. [32] performed a longitudinal study, providing long-term insights, while Sun [18] conducted a follow-up study to examine the continuity of effects. Most other studies featured short-term interventions or experiments, ranging from a few sessions to several weeks.

Table 1 Fundamental characteristics of the included studies

Studies	Year	Sample	Setting	Duration
[14]	2012	10 and 12 years old/fifth and seventh grade (30–35 students in each class)	Classroom	6 weeks
[15]	2012b	3rd grade (67 students)	Gym	6 weeks
[16]	2012	9–12 years old (4 classes: Avg. size 18–22 students)	Classroom	8 weeks
[17]	2013	4th grade (53 students)	Gym	9 months
[18]	2013	12 years old (70 students)	Classroom	6 weeks
[19]	2014	11–18 years old/ grades 6–12 (124 students)	Rooms	3 sessions of 60 min
[20]	2014	7–16 years old (20 students)	School	3 sessions of 180 min
[21]	2014	2nd-third grades (avg age: 8.18) (132 students)	Lab	6 weeks
[22]	2015	8–14 years old 24 unmotivated students	Exergame lab in the school	3 months
[23]	2015	11–18 years old/ grades 6–12 (124 students)	Rooms	3 sessions (60 min)
[24]	2015	13 years old/ sixth grade (11 students) and 18 university students	University campus	20 min
[25]	2015	4th, fifth grade or third, fourth grades (148 students)	Gym and lab	30 min
[26]	2015	6–7 years old (62 students)	Multipurpose room and playground	8 week
[27]	2016	10 years old (61 students)	Classroom	1 week
[28]	2016	Grades 4–6 (18 students)	Classroom	3 months
[29]	2016	7–12 years old (30 students)	University lab	30 min session
[30]	2017	Avg 8 years old/ second-third grade (261 students)	Classroom	2 weeks
[1]	2017	8–12 years old (345 students)	Classroom	6 weeks
[31]	2017	14–18 years old (37 overweight girls)	Research lab	12 weeks
[32]	2018	12–15 years old (190 students)	Classroom	8 weeks
[33]	2019	11–13 years old/seventh grade (52 students)	Classroom	2 weeks

(continued)

Table 1 (continued)

Studies	Year	Sample	Setting	Duration
[34]	2020	7–11 years old/ fourth, fifth grade (140 students)	Classrooms	3 sessions (120 min)
[35]	2020	Avg age 10 (53 students)	Classroom	8 weeks
[36]	2020	12 years old (140 students)	Classroom	2 sessions
[37]	2020	6th, 7th year (417 students)	Classroom	4 weeks
[38]	2020	8 teachers, 417 students (avg age 11.1, fifth, sixth grades)	Classroom	4–6 weeks
[39]	2020	3–5, 6–8, and 9–11 years old (131 students)	Classroom	2 sessions
[40]	2021	8–10 years old girls playing volleyball for 1 year+ (48 students)	Lab	8 weeks
[11]	2021	5th grade (120 students)	Gym	5 weeks
[41]	2022	Avg age: 7.3 (26 boys)	Gym	9 weeks
[42]	2022	8–10 years old girls (48 students)	School	6 weeks
[43]	2022	8–11 years old boys (30 students)	Online	9 weeks
[44]	2022	5th, 6th grade (34 students)	Classroom	12 weeks

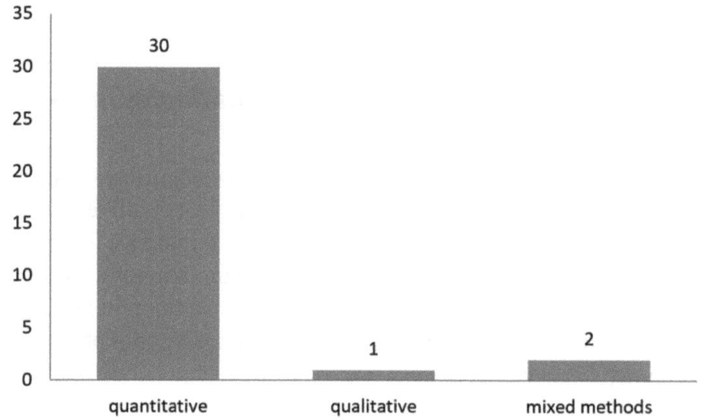

Fig. 3 AR and VR technologies used in PE in K-12 education

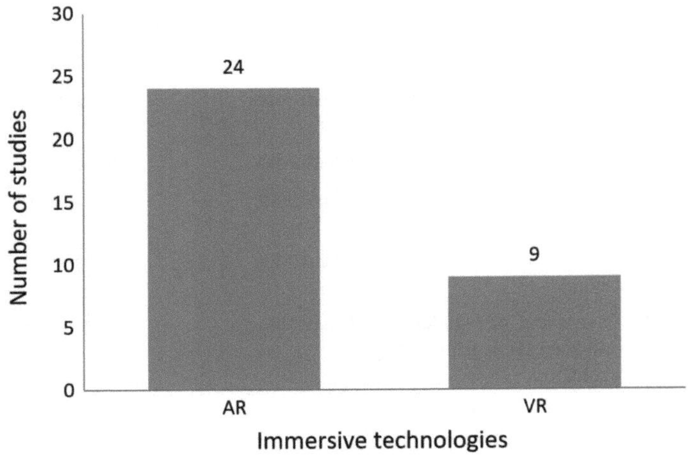

Fig. 4 AR and VR technologies used in PE in K-12 education

3.3 AR/VR Technology in K-12 PE

The integration of AR was the primary focus in most studies, whereas only nine studies specifically investigated the use of Virtual Reality (VR) in primary education PE, as illustrated in Fig. 4. Additionally, a significant proportion of the research, approximately 76%, concentrated on the potential of exergames, with 25 out of the total studies exploring their impact. This emphasis on exergames underscores their prominence in the application of immersive technologies within K-12 PE settings.

3.4 Benefits of the Use of Immersive Technologies in K-12 PE

The benefits found in the studies were put in conceptual categories, as illustrated in the tables below. Physical (Table 2), psychological (Table 3), cognitive (Table 4), social (Table 5), and attitudinal benefits (Table 6) of the use of immersive technologies in K-12 PE were found in the following articles. Each table summarizes the relevant studies that reported these specific types of benefits, providing a comprehensive overview of how AR and VR technologies positively impact various aspects of physical education for K-12 students.

3.4.1 Physical Benefits

Addressing RQ 1, the review identified several physical benefits of using immersive technologies in K-12 Physical Education (Table 2), as reported in 21 studies. The most frequently observed benefit was an increase in physical activity, noted in six studies

Table 2 Physical benefits

Physical benefits	No of studies ($N = 21$)	Studies
Increased physical activity	5	[18, 21, 22, 27, 31, 34]
Performance/efficient exercise	4	[14, 27, 30, 44]
Improve motor skills	2	[26, 33]
Movement skills/qualities	2	[20, 41]
Static and dynamic balance skills	2	[15, 35]
Physical fitness	1	[44]
Reduce of sedentary lifestyle	1	[43]
Effective in demonstrating human movement	1	[39]
Postural stability	1	[15]
Eye-hand coordination	1	[28]

Table 3 Psychological benefits

Psychological benefits	No of studies ($N = 26$)	Studies
Engagement/motivation	11	[1, 16, 19, 22–24, 27, 31, 33, 36, 38]
Enjoyment	6	[1, 18, 21, 25, 26, 41]
Improve mood	2	[34, 40]
Confidence/self-esteem/ self-efficacy	4	[11, 31, 34, 40]
Improve mental health satisfy psychological needs	1 1	[37, 40]
Reduce stress, anxiety, anger, mental confusion	1	[40]

Table 4 Cognitive benefits

Cognitive benefits	No of studies ($N = 7$)	Studies
Attention/concentration	3	[11, 20, 40]
Improve learning	1	[27]
In-depth study of content skills	1	[36]
Cognitive skills	1	[32]
Academic achievement	1	[37]

Table 5 Social benefits

Social benefits	Number of studies ($N = 6$)	Studies
Social skills	3	[14, 22, 32]
Collaboration	1	[22]
Group identification	1	[14]
Relatedness	1	[23]

Table 6 Attitudinal benefits

Attitudinal benefits	No of studies ($N = 10$)	Studies
Improve attitude toward physical activity	7	[11, 14, 17, 24, 31, 41, 43]
Positive attitude toward immersive technologies	3	[1, 38, 40]

[18, 21, 22, 27, 31, 34]. Enhanced performance and efficient exercise were reported in four studies [14, 27, 30, 44]. Improvements in motor skills were documented in two studies [26, 33], while other studies noted gains in movement skills and qualities [20, 41], static and dynamic balance skills [15, 35], and postural stability [15].

Additionally, individual studies highlighted specific benefits such as enhanced physical fitness [44], reduction in sedentary lifestyles [43], effective demonstration of human movement [39], improved eye-hand coordination [28], and better reaction times [40].

The physical benefits identified demonstrate that AR and VR technologies can effectively increase physical activity and improve various physical skills, making them valuable tools in PE curriculums. These technologies help students engage in more dynamic and interactive forms of exercise, contributing to their overall physical development and well-being.

3.4.2 Psychological Benefits

Regarding RQ3, Table 3 provides an overview of the psychological benefits observed, based on findings from 26 studies. The predominant psychological benefits included increased engagement and motivation, cited in 11 studies [1, 16, 19, 22–24, 27, 31, 33, 36, 38]. This suggests that AR and VR technologies can create more captivating and stimulating PE environments, which are crucial for sustaining students' interest and participation in physical activities.

Enhanced enjoyment was another significant benefit, reported in six studies [1, 18, 21, 25, 26, 41]. This indicates that immersive technologies can make physical education more enjoyable, potentially leading to higher levels of sustained physical activity and better adherence to fitness routines among students.

Immersive technologies have been shown to promote students emotional well-being [34, 40]. Four studies [11, 31, 34, 40] found that AR and VR can boost self-esteem, trust, and self-efficacy in physical activities, demonstrating their potential to empower students. Mental health benefits have also been reported, with research indicating that the usage of immersive technologies may decrease stress, anxiety, and mental confusion. Quintas et al. [37] also found that psychological needs such relatedness and autonomy can be satisfied.

3.4.3 Cognitive Benefits

Seven studies documented cognitive benefits (Table 4), such as improved learning, problem-solving, and academic achievement.

According to our review the most frequent cognitive benefit is increased attention and concentration as it was documented in [11, 20, 40]. These papers conclude that immersive technologies help maintain students' focus during PE activities, leading to improved cognitive performance.

Lindberg et al. [27] reported that AR and VR create engaging learning environments, helping students grasp complex concepts more effectively. Moreno-Guerrero et al. [36] found that these technologies facilitate a comprehensive exploration of educational material, catering to various learning styles and making content more accessible.

Ruiz-Ariza et al. [32] observed improvements in problem-solving and critical thinking due to the interactive nature of AR and VR. Quintas et al. [37] noted that the use of immersive technologies can enhance academic achievement, likely due to increased engagement and motivation.

3.4.4 Social Benefits

AR and VR were also found to enhance students' social skills and collaboration (Table 5). Several studies [14, 22, 32] have reported that these technologies improve students' ability to communicate and collaborate with peers effectively. Increased collaboration was highlighted by Finco et al. [22] who discovered that AR and VR experiences often require teamwork and cooperative learning, thereby cultivating essential social skills. Additionally, [14] observed that immersive technologies facilitate group identification and foster a sense of belonging, thereby enabling students to feel more connected to their peers and nurturing a supportive community within the context of physical education. Kooiman and Sheehan [23] also reported that these technologies increase students' sense of relatedness, leading to a more positive and collaborative classroom atmosphere. This improved sense of connection with peers not only promotes social development but also contributes to the creation of a more inclusive and engaging learning environment.

The social benefits outlined in Table 5 illustrate how AR and VR technologies may greatly improve students' social skills, collaborative abilities, and sense of community.

3.4.5 Attitudinal Benefits

Regarding the attitudinal benefits (RQ5) of using immersive technologies in K-12 PE, this review identified several key findings. Immersive technologies have been shown to positively influence students' attitudes toward physical activity and the use of technology in education.

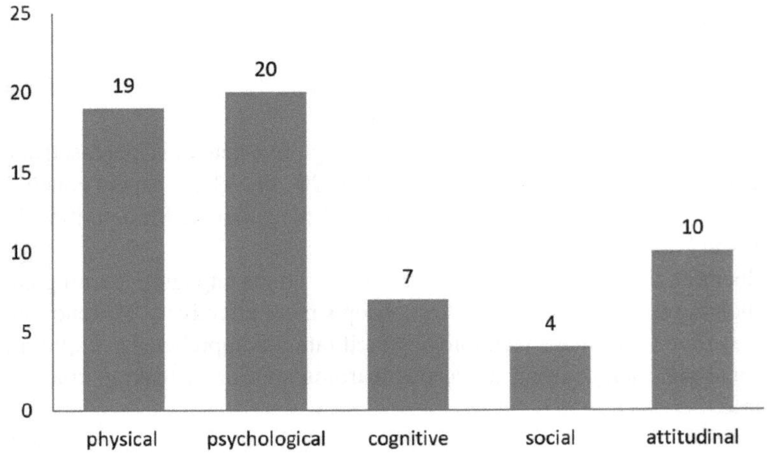

Fig. 5 Benefits of the use of immersive technologies in K-12 PE education

Studies reported in Table 6 indicate that AR and VR significantly improve students' attitudes toward physical activity. Seven studies [11, 14, 17, 24, 31, 41, 43] found that students engaged with these technologies are more enthusiastic about participating in physical activities, suggesting a shift toward a more active lifestyle. Furthermore, these technologies foster positive attitudes toward the use of immersive technologies themselves [1, 38, 40] noted that students showed increased interest and positive perceptions of AR and VR in educational settings. This positive reception is crucial for the successful integration of new technologies in the classroom, as it can lead to higher levels of engagement and motivation.

Overall, the majority of the studies explored psychological and physical benefits, while fewer studies addressing attitudinal, cognitive and social benefits in descending order, as illustrated in Fig. 5.

While most studies reported positive outcomes, Table 7 summarizes the negative or null effects. For example, Shewmake et al. [25] found that some students felt they were working less intensely during immersive technology-based activities. Sun [18] observed lower levels of physical activity in certain contexts when using these technologies. Wadsworth et al. noted reduced participation rates in some instances [21]. Additionally, Pedersen [29] reported no significant effects on lateral movement skills. These findings suggest that while immersive technologies offer many benefits, their implementation should be carefully considered to avoid potential drawbacks and ensure they complement traditional PE effectively.

Table 7 Zero or negative effects

Zero or negative effects	Studies
Feeling of working less	[25]
Lower physical activity	[18]
Lower participation	[21]
No effects on lateral movement	[29]

4 Discussion

This systematic literature review examined how immersive technologies—specifically, AR and VR—affect physical education in grades K-12. The physical, cognitive, psychological, and social dimensions of student engagement and learning in physical education are all improved by immersive technology, according to consistent data found in this research.

The findings reveal that by offering a fun substitute for traditional physical education approaches, AR/VR technologies assist raise students' levels of physical activity and enhance their motor abilities. According to studies, the interactive and gamified feature of AR/VR settings motivates students to practice physical skills more readily [4, 30]. Additionally, AR applications may show students how to perform exercises correctly, enabling them to self-correct and improve their movements. It has been proven that this feature improves workout performance.

Significant physical benefits, such as increased physical activity, improved motor skills, and improved balance and coordination, were identified in the review. For instance, Lindberg et al. and Andrade et al. found that AR and VR greatly boost students' levels of physical activity [27, 34]. Immersive technologies have been shown to psychologically increase student motivation, engagement, and enjoyment in physical education classes [1, 22, 33]. These findings are significant because they imply that AR and VR may successfully counteract sedentary lifestyles and support students' physical and mental well-being.

Additionally, there were notable cognitive benefits such improved attention, concentration, and problem-solving abilities. Research [32, 36] shows how these technologies enhance dynamic learning settings, which improve cognitive growth and academic achievement. Attitudinal benefits were evident, with students developing more positive attitudes toward physical activity and the use of technology in education [41, 43]. This shift in attitude is vital for the acceptance and integration of new technologies in educational curricula.

Socially, immersive technologies enhance students' ability to communicate and collaborate with peers. The interactive and engaging nature of AR and VR fosters teamwork and cooperative learning, essential skills for students' overall development [14, 22]. Enhanced social interactions and a sense of community within PE settings suggest that these technologies can create more inclusive and supportive learning environments [23].

In contrast, there are some studies, in which negative or no effects have been found, suggesting that immersive technologies in PE should be used with caution or as a supplementary resource [16, 29]. In some cases, it has been observed that students' engagement with immersive technologies lead to lower physical activity [18] or lower participation [21]. while no effects were observed on lateral movement [29] when compared to regular PE. It was also reported that students had the feeling of working less, although this feeling might have emerged because they enjoyed exergaming to such a degree that they did not realize their effort [25].

5 Conclusion

This systematic review highlights the significant potential of immersive technologies, to enhance K-12 Physical Education. The findings indicate substantial benefits across physical, psychological, cognitive, social, and attitudinal domains, demonstrating that these technologies can transform traditional PE practices into more engaging and effective learning experiences.

Despite the promising findings, existing research still has notable gaps. Little research has been done on the long-term consequences of integrating AR and VR in PE settings, with most studies concentrating on short-term benefits. Furthermore, future study on immersive technology in PE has to include a wider range of research approaches. Students' subjective experiences and impressions are not fully understood due to the prevalent adoption of quantitative methodologies. The qualitative studies that may offer more in-depth understanding of how teachers and students interact with these devices are conspicuously lacking. Nevertheless, using mixed methods research would provide deeper, more complex insights into the ways that AR and VR impact students, capturing both quantifiable results and individual experiences. In the end, a greater comprehension of the holistic benefits of immersive technologies in education can result from the use of mixed methodologies to assist bridge the gap between numerical data and students' real-world experiences. Furthermore, studies with good results may have a higher chance of being published than those with null or negative results, which raises the possibility of publication bias.

The adoption of immersive technologies in PE faces several challenges, including limited resources, lack of teacher training, technical issues, and perceived increased workload. Teachers frequently lack requisite training, and many schools lack dependable infrastructure and AR/VR devices, which causes opposition or inappropriate usage of immersive technologies. To overcome these obstacles and guarantee that immersive technologies are successfully incorporated into physical education that benefit both students and teachers, specific professional development and infrastructure support, such as funding for equipment and improved internet connectivity are required.

The effective use of AR and VR in physical education curriculums requires careful assessment of the available resources and instructor readiness. A smart place to start is with simple AR applications that support marker-based AR activities that use

QR codes to present relevant multimedia content during physical activities. This approach is cost-effective and easily accessible, making it suitable even for schools with limited resources [3]. There are additional options for picture augmentation that do not depend on QR codes, making the process more engaging and user-friendly. For instance, ARTutor [45] is a free, education-oriented AR application that simplifies implementation of augmentation experiences. It is consisted of a web-based authoring tool and a mobile application for AR content presentation. It lets educators easily add augmentations over images without changing the original content, offering a handy and easy tool for immersive learning experiences. Another effective strategy is leveraging exergames, which need little extra equipment but can rally boost student interest and action levels [5]. These activities can be blended into standard PE lessons to offer a mix of physical activity and interactive learning, increasing engagement.

Educators need adequate training in how to use these technologies effectively. Training programs that emphasize incorporating immersive technologies into typical instruction can enhance teachers confidence and reduce the perceived cognitive overload associated with these technologies tools [27].

The current systematic review also has several limitations that need to be addressed. First, the review data collection was limited to indexed articles only in Google Scholar, which may exclude relevant research indexed in other databases. Furthermore, research published in conference proceedings and collective volumes of books was omitted. Most studies also focus solely on certain advantages such as motivation and engagement, ignoring others such as emotional well-being and the long-term influence on students' attitudes toward physical exercise.

Future research should address these limitations by increasing the sample of the review to research from diverse sources (e.g., conference papers and book chapters) to broaden their coverage of the research undertaken on the topic. Further research should systematically investigate the long-term effects and the emotional impact on students' attitudes. Additionally, research on the cost-effectiveness and accessibility of these technologies will be crucial for widespread adoption. Finally studies should investigate the best practices for integrating AR and VR into PE curricula, considering various educational contexts and diverse student populations.

Disclosure of Interests The authors have no competing interests to declare that are relevant to the content of this article.

References

1. Ho, S.S., Lwin, M.O., Sng, J.R.H., Yee, A.Z.H.: Escaping through exergames: presence, enjoyment, and mood experience in predicting children's attitude toward exergames. Comput. Hum. Behav. **72**, 381–389 (2017). https://doi.org/10.1016/j.chb.2017.03.001
2. Benzing, V., Schmidt, M.: Exergaming for children and adolescents: strengths, weaknesses, opportunities and threats. J. Clin. Med. **7**, 422 (2018). https://doi.org/10.3390/jcm7110422

3. Gao, Z.: Fight fire with fire? Promoting physical activity and health through active video games. J. Sport Health Sci. **6**, 1–3 (2017). https://doi.org/10.1016/j.jshs.2016.11.009
4. Calabuig-Moreno, F., González-Serrano, M.H., Fombona, J., García-Tascón, M.: The emergence of Technology in physical education: a general bibliometric analysis with a focus on virtual and augmented reality. Sustainability. **12**, 2728 (2020). https://doi.org/10.3390/su12072728
5. Suriya, P.: Arumugam. **9**, 9413–9416 (2020)
6. Gao, Z., Chen, S., Pasco, D., Pope, Z.: A meta-analysis of active video games on health outcomes among children and adolescents. Obes. Rev. **16**, 783–794 (2015). https://doi.org/10.1111/obr.12287
7. Maas, M.J., Hughes, J.M.: Virtual, augmented and mixed reality in K-12 education: a review of the literature. Technol. Pedag. Educ. **29**, 231–249 (2020). https://doi.org/10.1080/1475939X.2020.1737210
8. Papanastasiou, G., Drigas, A., Skianis, C., Lytras, M., Papanastasiou, E.: Virtual and augmented reality effects on K-12, higher and tertiary education students' twenty-first century skills. Virtual Real. **23**, 425–436 (2019). https://doi.org/10.1007/s10055-018-0363-2
9. Ariffin, U.H.B., Mohamed Mokmin, N.A.B.: Augmented reality technology in physical education: a systematic review in instructional design, and AR implementation option over the last 5 years. Adv. J. Tech. Vocat. Educ. **6**, 13–20 (2022). https://doi.org/10.26666/rmp.ajtve.2022.1.3
10. Ennis, C.D.: Implications of exergaming for the physical education curriculum in the 21st century. J. Sport Health Sci. **2**, 152–157 (2013). https://doi.org/10.1016/j.jshs.2013.02.004
11. Lee, H.S., Lee, J.: The effect of elementary school soccer instruction using virtual reality technologies on students' attitudes toward physical education and flow in class. Sustainability. **13**, 3240 (2021). https://doi.org/10.3390/su13063240
12. Sargent, J., Calderón, A.: Technology-enhanced learning physical education? A critical review of the literature. J. Teach. Phys. Educ. **41**, 689–709 (2021). https://doi.org/10.1123/jtpe.2021-0136
13. Yang, Q.-F., Hwang, G.-J., Sung, H.-Y.: Trends and research issues of mobile learning studies in physical education: a review of academic journal publications. Interact. Learn. Environ. **28**, 419–437 (2020). https://doi.org/10.1080/10494820.2018.1533478
14. Lwin, M.O., Malik, S.: The efficacy of exergames-incorporated physical education lessons in influencing drivers of physical activity: a comparison of children and pre-adolescents. Psychol. Sport Exerc. **13**, 756–760 (2012). https://doi.org/10.1016/j.psychsport.2012.04.013
15. Sheehan, D.P., Katz, L.: The effects of a daily, 6-week exergaming curriculum on balance in fourth grade children. J. Sport Health Sci. **2**, 131–137 (2013). https://doi.org/10.1016/j.jshs.2013.02.002
16. Sun, H.: Exergaming impact on physical activity and interest in elementary school children. Res. Q. Exerc. Sport. **83**, 212–220 (2012). https://doi.org/10.1080/02701367.2012.10599852
17. Gao, Z., Zhang, T., Stodden, D.: Children's physical activity levels and psychological correlates in interactive dance versus aerobic dance. J. Sport Health Sci. **2**, 146–151 (2013). https://doi.org/10.1016/j.jshs.2013.01.005
18. Sun, H.: Impact of exergames on physical activity and motivation in elementary school students: a follow-up study. J. Sport Health Sci. **2**, 138–145 (2013). https://doi.org/10.1016/j.jshs.2013.02.003
19. Kooiman, B.J., Sheehan, D.P.: Motivation to move with exergaming in online physical education. Int. J. Phys. Educ. Fit. Sports. **3**, 01–24 (2014). https://doi.org/10.26524/1421
20. Meckbach, J., Gibbs, B., Almqvist, J., Quennerstedt, M.: Wii teach movement qualities in physical education. Sport Sci. Rev. **23**, 241–266 (2014). https://doi.org/10.1515/ssr-2015-0004
21. Wadsworth, D., Brock, S., Daly, C., Robinson, L.: Elementary students' physical activity and enjoyment during active video gaming and a modified tennis activity. J. Phys. Educ. Sport. **14**, 311 (2014). https://doi.org/10.7752/jpes.2014.03047
22. Finco, M.D., Reategui, E., Zaro, M.A., Sheehan, D.D., Katz, L.: Exergaming as an alternative for students unmotivated to participate in regular physical education classes. Int. J. Game-Based Learn. **5**, 1–10 (2015). https://doi.org/10.4018/IJGBL.2015070101

23. Kooiman, B.J., Sheehan, D.P.: The efficacy of exergames for social relatedness in online physical education. Cogent Educ. **2**, 1045808 (2015). https://doi.org/10.1080/2331186X.2015.1045808
24. Laine, T.H., Suk, H.J.: Designing mobile augmented reality exergames. Games Cult. **11**, 548–580 (2016). https://doi.org/10.1177/1555412015572006
25. Shewmake, C.J., Merrie, M.D., Calleja, P.: Xbox Kinect gaming systems as a supplemental tool within a physical education setting: third and fourth grade students' perspectives. Phys. Educ. (2015). https://doi.org/10.18666/TPE-2015-V72-I5-5526
26. Vernadakis, N., Papastergiou, M., Zetou, E., Antoniou, P.: The impact of an exergame-based intervention on children's fundamental motor skills. Comput. Educ. **83**, 90–102 (2015). https://doi.org/10.1016/j.compedu.2015.01.001
27. Lindberg, R., Seo, J., Laine, T.H.: Enhancing physical education with Exergames and wearable technology. IEEE Trans. Learn. Technol. **9**, 328–341 (2016). https://doi.org/10.1109/TLT.2016.2556671
28. Ma, A.W.W., Qu, L.: The effect of exergaming on eye-hand coordination among primary school children: a pilot study. Adv. Phys. Educ. **6**, 99–102 (2016). https://doi.org/10.4236/ape.2016.62011
29. Pedersen, S.J., Cooley, P.D., Cruickshank, V.J.: Caution regarding exergames: a skill acquisition perspective. Phys. Educ. Sport Pedag. **22**, 246–256 (2017). https://doi.org/10.1080/17408989.2016.1176131
30. Gao, Z., Pope, Z., Lee, J.E., Stodden, D., Roncesvalles, N., Pasco, D., Huang, C.C., Feng, D.: Impact of exergaming on young children's school day energy expenditure and moderate-to-vigorous physical activity levels. J. Sport Health Sci. **6**, 11–16 (2017). https://doi.org/10.1016/j.jshs.2016.11.008
31. Staiano, A.E., Beyl, R.A., Hsia, D.S., Katzmarzyk, P.T., Newton, R.L.: Twelve weeks of dance exergaming in overweight and obese adolescent girls: transfer effects on physical activity, screen time, and self-efficacy. J. Sport Health Sci. **6**, 4–10 (2017). https://doi.org/10.1016/j.jshs.2016.11.005
32. Ruiz-Ariza, A., Casuso, R.A., Suarez-Manzano, S., Martínez-López, E.J.: Effect of augmented reality game Pokémon GO on cognitive performance and emotional intelligence in adolescent young. Comput. Educ. **116**, 49–63 (2018). https://doi.org/10.1016/j.compedu.2017.09.002
33. Chang, K.-E., Zhang, J., Huang, Y.-S., Liu, T.-C., Sung, Y.-T.: Applying augmented reality in physical education on motor skills learning. Interact. Learn. Environ. **28**, 685–697 (2020). https://doi.org/10.1080/10494820.2019.1636073
34. Andrade, A., Cruz, W.M.D., Correia, C.K., Santos, A.L.G., Bevilacqua, G.G.: Effect of practice exergames on the mood states and self-esteem of elementary school boys and girls during physical education classes: a cluster-randomized controlled natural experiment. PLoS One. **15**, e0232392 (2020). https://doi.org/10.1371/journal.pone.0232392
35. Demir, A., Akin, M.: The effect of Exergame education on balance in children. Malays. Online J. Educ. Technol. **8**, 100–107 (2020). https://doi.org/10.17220/mojet.2020.03.006
36. Moreno-Guerrero, A.-J., Alonso García, S., Ramos Navas-Parejo, M., Campos-Soto, M.N., Gómez García, G.: Augmented reality as a resource for improving learning in the physical education classroom. Int. J. Environ. Res. Public Health. **17**, 3637 (2020). https://doi.org/10.3390/ijerph17103637
37. Quintas, A., Bustamante, J.-C., Pradas, F., Castellar, C.: Psychological effects of gamified didactics with exergames in physical education at primary schools: results from a natural experiment. Comput. Educ. **152**, 103874 (2020). https://doi.org/10.1016/j.compedu.2020.103874
38. Quintas-Hijós, A., Peñarrubia-Lozano, C., Bustamante, J.C.: Analysis of the applicability and utility of a gamified didactics with exergames at primary schools: qualitative findings from a natural experiment. PLoS One. **15**, e0231269 (2020). https://doi.org/10.1371/journal.pone.0231269
39. Spinosa, R.M.D.O., Santo, D.L.D., Costa, R.Z.F., Pazetto, N.F., Santos, C.R.D., Medina-Papst, J., Marques, I.: Comparing live and digital augmented reality models for demonstrating two motor skills from the test of gross motor development—second edition: TGMD-2. Percept. Mot. Skills. **127**, 386–400 (2020). https://doi.org/10.1177/0031512519900061

40. Amprasi, E., Vernadakis, Z., Zetou, E., Antoniou, P.: The impact of a full immersive virtual reality intervention on children's reaction time. PANR J. (2021). https://www.panr.com.cy/?p=7412
41. Juliantine, T., Setiawan, E., Jumareng, H., Gani, R.A., Asnaldi, A.: Do fundamental movement skills, physical activity and enjoyment among inactive student during the COVID-19 era improve after exergame? J. Phys. Educ. **33** (2022). https://doi.org/10.4025/jphyseduc.v33i1.3327
42. Amprasi, E., Vernadakis, N., Zetou, E., Antoniou, P.: Effect of a full immersive virtual reality intervention on selective attention in children. Int. J. Instr. **15**, 565–582 (2022). https://doi.org/10.29333/iji.2022.15132a43
43. Bisgin, H.: The impact of virtual physical education on physical activity and the reduction of sedentary of elementary students in the Covid-19 era. Pak. J. Med. Health Sci. **16**, 825–832 (2022). https://doi.org/10.53350/pjmhs22162825
44. Ketelhut, S., Röglin, L., Martin-Niedecken, A.L., Nigg, C.R., Ketelhut, K.: Integrating regular exergaming sessions in the ExerCube into a school setting increases physical fitness in elementary school children: a randomized controlled trial. J. Clin. Med. **11**, 1570 (2022). https://doi.org/10.3390/jcm11061570
45. Terzopoulos, G., Kazanidis, I., Tsinakos, A.: Building a general purpose educational augmented reality application: the case of ARTutor. In: Auer, M.E., Tsiatsos, T. (eds.) New Realities, Mobile Systems and Applications, pp. 168–179. Springer International Publishing, Cham (2022)

MIX
Papier aus verantwortungsvollen Quellen
Paper from responsible sources
FSC® C105338

If you have any concerns about our products,
you can contact us on
ProductSafety@springernature.com

In case Publisher is established outside the EU,
the EU authorized representative is:
Springer Nature Customer Service Center GmbH
Europaplatz 3, 69115 Heidelberg, Germany

Printed by Libri Plureos GmbH
in Hamburg, Germany